T0134618

Springer Series in Measurement Science and Technology

The Springer Series in Measurement Science and Technology comprehensively covers the science and technology of measurement, addressing all aspects of the subject from the fundamental principles through to the state-of-the-art in applied and industrial metrology, as well as in the social sciences. Volumes published in the series cover theoretical developments, experimental techniques and measurement best practice, devices and technology, data analysis, uncertainty, and standards, with application to physics, chemistry, materials science, engineering and the life and social sciences.

More information about this series at http://www.springer.com/series/13337

Michael Quinten

A Practical Guide to Surface Metrology

Michael Quinten
Aldenhoven, Germany

ISSN 2198-7807 ISSN 2198-7815 (electronic)
Springer Series in Measurement Science and Technology
ISBN 978-3-030-29456-4 ISBN 978-3-030-29454-0 (eBook)
https://doi.org/10.1007/978-3-030-29454-0

This Springer imprint is published by the registered company Springer Nature Switzerland AG.
The registered company address is: Gewerbestrasse 11, 6330 Cham, Switzerland

To my family

Preface

You must measure the things that are measurable, and make measurable the things that are not.
Galileo Galilei (1564–1642)

New and improved materials, better machines, and phenomenal progress in manufacturing contribute to technological challenges of today for inspection methods and techniques. In the production of mechanical, electronic, or optical microcomponents, the dimensions are even continuously scaling down, while the surface structures increasingly become more complex. In consequence, development, process control, and quality control must be carried out on these size scales with sophisticated measuring devices.

Only with metrology we are able to check our manufacturing, to describe our product, to establish models for our planning, and to validate the models. The corresponding applied metrology tools must then meet the specific requirements. Hence, to keep pace with the demands from industry for greater accuracy, repeatability, reliability, ease of use, and continuous process control, new and improved measuring techniques were developed.

The purpose of this book is to introduce the commonly used measuring techniques of today. In doing so, the book deals with many, although by no means all, methods of surface metrology. Some of them already have a long tradition but have experienced big progress not only with the progress in the computer capabilities required to handle large amounts of data but also with the development of intense light sources and more sensitive detectors. The biggest part of measuring methods is related with optical methods and imaging, especially those which are basing on interferometry. But also tactile, capacitive, and inductive methods are considered.

There are other books on dimensional metrology that cover many of the methods presented in this book. While *Optical Shop Testing* (2007) edited by Daniel Malacara concentrates on almost all optical methods for examination of optical elements and their surface, the book *Optical Measurement of Surface Topography* (2011) edited by Richard Leach gives a comprehensive overview on optical methods for dimensional surface topography measurement at that time. The *Handbook of*

Optical Metrology: Principles and Applications edited by Tōru Yoshizawa and revised in 2015 is quite similar to the book of Richard Leach, with contributions of other experts in the field. The book *Characterisation of Areal Surface Texture* (2013) edited by Richard Leach concentrates on characterization techniques particularly for areal surface texture.

Books that emphasize more thc scientific aspects of surfaces and interfaces and the corresponding measuring methods are also available. *Surface Science Techniques* (2013) edited by Gianangelo Bracco and Bodil Holst describes the experimental techniques employed to study surfaces and interfaces in which the emphasis is laid on methods and techniques used in surface science. *Surface and Interface Science* (2012) with six volumes edited by Klaus Wandelt covers interface science from a new surface science perspective and is a rather good reference for scientific people but less for practitioners.

To write this book finally required reading and valuating many monographs as such mentioned above and a still larger number of publications related to dimensional surface metrology. The total amount of published work is, however, too immense to consider it all. Therefore, I hope to have included the most relevant up to date and apologize for all the contributions not considered here.

As the book provides access to many principles and practices of modern dimensional measurement on surfaces, I hope that this book will prove of value to developers and practitioners in the field of dimensional surface metrology. Nevertheless, also graduated students and lecturers may benefit from the information contained in this book on the various measuring techniques.

In the end, I want to express my thanks to some people. First of all, many reputable manufacturers of measuring instruments were helpful in supplying photographic material and measurement results for illustrating. Many thanks for their cooperation! I also would like to acknowledge the continuing support of Tom Spicer and Hisako Niko from Springer Nature. They were the "helping hand" in all questions concerning publishing this book. Last but not least, I want to thank my family for the support and patience with me during writing this book.

Aldenhoven, Germany Michael Quinten
June 1, 2019

Contents

1 Introduction to Surfaces and Surface Metrology 1
1.1 Microscopic View on a Surface 3
1.2 Macroscopic View on a Surface 5
1.3 Measurement and Validation 8
1.3.1 Profile Measurement 16
1.3.2 Areal Measurement 27
1.3.3 Measurement Modules 31
1.3.4 The Way to Reliable Surface Data 39
References 41

2 Tactile Surface Metrology 43
2.1 Tactile Surface Profiling 43
2.2 Atomic Force Microscopy 47
References 54

3 Capacitive and Inductive Surface Metrology 57
3.1 Capacitive Surface Profiling 57
3.2 Surface Profiling with Eddy Currents 62
References 65

4 Optical Surface Metrology – Physical Basics 67
4.1 Electromagnetic Waves 67
4.2 Huygens-Fresnel Principle of Wave Propagation 70
4.3 Polarization 71
4.4 Interference 72
4.5 Coherence 74
4.6 Dielectric Function and Refractive Index 75
4.7 Reflection and Refraction 81
4.8 Dispersion Effects 84

4.9 Diffraction 86
4.10 Scattering 88
References 92

5 **Optical Surface Metrology: Methods** 95
5.1 Chromatic Confocal Surface Profiling 95
5.2 Surface Profiling with an Autofocus Sensor 102
5.3 Light Sectional Methods 104
5.3.1 Laser Point Triangulator 104
5.3.2 Line Projection 105
5.3.3 Fringe Projection 106
5.4 Microscopy Methods 108
5.4.1 Classical Microscopy 109
5.4.2 Confocal Microscopy 115
5.4.3 Focal Depth Variation 124
5.4.4 Scanning Near-Field Optical Microscopy 127
5.5 Interferometric Methods 128
5.5.1 Interferometric Form Inspection 132
5.5.2 Tilted Wave Interferometry 134
5.5.3 White Light Interferometry 136
5.5.4 Wavelength Scanning Interferometry 145
5.5.5 Multi-Wavelength Interferometry 147
5.5.6 Grazing Incidence Interferometry 150
5.5.7 Frequency Scanning Interferometry 153
5.5.8 Digital Holographic Microscopy 155
5.5.9 Conoscopy 159
5.6 Wave Front Sensing (Shack-Hartmann) 160
5.7 Deflectometry 163
5.8 Makyoh Topography Sensor 165
5.9 Surface Profiling Using Elastic Light Scattering 167
5.9.1 Total Integrated Scattering (TIS) 168
5.9.2 Angular Resolved Scattering (ARS) 169
5.9.3 Speckle Based Roughness Determination 173
5.10 Spectral Analysis and Characterization 176
5.10.1 Reflectometry 177
5.10.2 Spectroscopic Ellipsometry 188
References 190

6 **Imaging Methods** 199
6.1 Industrial Image Processing 199
6.1.1 Illumination 201
6.1.2 Camera 201
6.1.3 Image Processing Hardware 202
6.1.4 Digital Image Processing Software 202
6.1.5 Mechanics 203

6.2 Shape from Shading 203
6.3 Hyperspectral Imaging 205
6.4 Scanning Electron Microscopy 208
6.5 Optical Coherence Tomography 211
6.6 Terahertz Spectroscopy 213
References 216

7 Multisensor – Systems – A Versatile Approach to Surface Metrology 219

Appendix – Numerics with Complex Numbers 223

Index 227

About the Author

Michael Quinten During his academic period from 1983 to 2000, Michael Quinten obtained his diploma degree and his Ph.D. in Physics at the University of Saarland, Saarbruecken, Germany, and joined in 1990 the RWTH Aachen University for habilitation. After the successful habilitation, he spent 4 years at several universities in Graz (Austria), Chemnitz, Aachen, Saarbruecken, and Bochum. Altogether, he authored in this academic period more than 50 scientific publications with topics in optical properties of nanoparticles, nanoparticle materials, and aerosols.

In 2001, he joined the STEAG ETA-Optik GmbH, Germany, where he first worked as head of R&D of integrated optics components and later became product manager in the Colour and Coatings Division. In 2007, he moved to FRT GmbH, Germany, where he presently works as head of R&D Sensors and is responsible for the optical sensor technology division.

Already in 1996, he also founded *Dr. Michael Quinten – Wissenschaftlich Technische Software* where he develops and retails highly sophisticated scientific software. The software results from the long-lasting practical and theoretical experiences of him in solid-state physics, electrodynamics, and the analysis of layer thickness and colour.

In 2011, he published his first book *Optical Properties of Nanoparticle Systems: Mie and Beyond*, a comprehensive summary of his work in the academic period. Already in 2012 followed the second book *A Practical Guide to Optical Metrology for Thin Films*, proving his expertise in optical thin film measurement. The third book *Optische Schichtdickenmessung mit miniaturisierten Spektrometern* in 2015 is the German counterpart to the second book. Finally, in 2018 appeared the fourth book *Practical Determination of Optical Constants from Spectral Measurements*, a short guide to the determination of optical constants.

List of Figures

Fig. 1.1 Normalized charge densities of ions and free electrons at a metal surface according to the jellium model 4
Fig. 1.2 Illustrations of (**a**) the nominal surface, (**b**) the real surface, (**c**) the measured surface 6
Fig. 1.3 Division of shape deviations 7
Fig. 1.4 Illustration of the lateral resolution in microscopy according to the Rayleigh criterion, the Abbe criterion, and the Sparrow criterion for a wavelength of $\lambda = 560$ nm and a numerical aperture of $NA = 0.8$ 11
Fig. 1.5 Illustration of the definition of accuracy 11
Fig. 1.6 Examples of certified standards: (**a**) depth setting standards with round and rectangular grooves, (**b**) rougness measurement standards, (**c**) ceramic end-gauge blocks, (**d**) flatness standard (optical flat), (**e**) angle standards. Pictures for (**a** and **b**) courtesy of HALLE Präzisions-Kalibriernormale GmbH, Edemissen, Germany, for (**c** and **d**) courtesy of Mitutoyo Deutschland GmbH, Neuss, Germany, www.mitutoyo.de, and for (**e**) courtesy of SARTORIUS Werkzeuge GmbH & Co. KG, Ratingen, Germany 12
Fig. 1.7 Effects that influence the measured profile. For description see text 17
Fig. 1.8 Primary profile with the tracing length divided into an evaluation length, a pre travel length, and a post travel length 18
Fig. 1.9 Transfer characteristic and mode of operation of the noise filter, roughness filter, and form filter 19
Fig. 1.10 Separated roughness profile from the primary profile in Fig. 1.8 20
Fig. 1.11 Separated waviness profile from the primary profile in Fig. 1.8 20

Fig. 1.12 Illustrations of (**a**) an exemplaric roughness profile, (**b**) its height distribution histogram, (**c**) the parameter R_a, (**d**) the parameter R_q 21
Fig. 1.13 Illustration of the parameters R_v, R_p, and R_t 23
Fig. 1.14 Illustration of how to determine the parameter R_z 23
Fig. 1.15 Indication of surface texture according to ISO 1302 25
Fig. 1.16 Control elements of surface texture requirement 26
Fig. 1.17 Definition of the S-F surface and the S-L surface according to ISO 25178-2 30
Fig. 1.18 Indication of surface texture according to ISO 25178-1 30
Fig. 1.19 Control elements of areal surface texture requirement 31
Fig. 1.20 Response curve of a semiconductor photodiode in dependence on the amount of light *H* striking the photodiode 36
Fig. 1.21 Quantum efficiency of a silicon CMOS image detector in dependence on the wavelength of incident radiation. The oscillations are caused by a thin transparent protective quartz window in front of the detector 37
Fig. 1.22 Quantum efficiency of a front-illuminated CCD image sensor, a back-thinned CCD image sensor, and a back-thinned CCD linear detector in dependence on the wavelength of incident light 37
Fig. 1.23 Photosensitivity of a linear InGaAs detector array in dependence on the wavelength of the incident radiation 38
Fig. 1.24 Typical spectra of a white light LED, a cyan LED (center wavelength 520 nm) and two superluminescence diodes with their center wavelength in the near infrared at 830 nm and 1310 nm 39

Fig. 2.1 Sketch of the working principle of a stylus tip 44
Fig. 2.2 Typical stylus tip with spherical shape used in coordinate measuring machines (CMM). (Courtesy of Goekeler Messtechnik GmbH, Lenningen, Germany) 44
Fig. 2.3 Influence of the probing tip size on the measured profile. On bottom the original profile. The measured profiles are shifted along the ordinate by a constant value for better presentation 45
Fig. 2.4 Typical high precision stylus measurement system (left) and sensing arm for roughness measurement (right). (Courtesy of AMETEK GmbH – Business Unit Taylor Hobson, Weiterstadt, Germany) 46
Fig. 2.5 Surface roughness mappings of two different surfaces with a stylus tip. (Courtesy of AMETEK GmbH – Business Unit Taylor Hobson, Weiterstadt, Germany) 47
Fig. 2.6 Sketch of the principle of an atomic force microscope (AFM) 47
Fig. 2.7 AFM pictures (**a**) compact system CoreAFM, (**b**) lens system LensAFM. (Courtesy of Nanosurf AG, Liestal, Switzerland) 48

Fig. 2.8 Pictures of typical AFM cantilevers with tips. (© 2017 NanoWorld AG, Neuchâtel, Switzerland) 49
Fig. 2.9 AFM topography measurements: (**a**) two fingers of a photonic band gap filter, (**b**) surface topography of a sapphire wafer with PSS stuctures. Courtesy of FRT GmbH, Bergisch Gladbach, Germany 52

Fig. 3.1 (**a**) Empty capacitor, (**b**) Capacitor filled with a dielectric material 58
Fig. 3.2 Sketch of a shielded measuring electrode in a probe head as used in surface metrology 58
Fig. 3.3 (**a**) Standard capacitive sensor types, (**b**) World's smallest capacitive sensor. (Pictures are courtesy of E + H Metrology GmbH, Karlsruhe, Germany, www.eh-metrology.com) 59
Fig. 3.4 Dual capacitive sensor setup 60
Fig. 3.5 Capacitive surface mapping of a silicon wafer: **a** topography from top, **b** topography from bottom, **c** the warp of the wafer. (Courtesy of E + H Metrology GmbH, Karlsruhe, Germany, www.eh-metrology.com) 61
Fig. 3.6 Working principle of an eddy current sensor 62
Fig. 3.7 The normalized impedance plane of an eddy current sensor. The black point defines the working point. The blue line indicates the undisturbed dependence of the coil impedance upon the conductivity. Deviations from the blue line are indicated in red. For their description see text 64

Fig. 4.1 The complete spectral range of electromagnetic radiation. The colored region from 380 to 780 nm is the visible spectral range 68
Fig. 4.2 Sketch of a harmonic electromagnetic wave 69
Fig. 4.3 Huygens' principle applied on a plane wave 71
Fig. 4.4 Sketch of (**a**) a linearly polarized wave, (**b**) a circularly polarized wave, and (**c**) an elliptically polarized wave 73
Fig. 4.5 Illustration of (**a**) arbitrary interference, (**b**) destructive interference, and (**c**) constructive interference of two waves with same magnitude 74
Fig. 4.6 Natural light as Gaussian wave packet with center wavelength λ_0 74
Fig. 4.7 Dielectric function and refractive index of a harmonic oscillator with oscillator strength $S = 1$, resonance frequency $3.5 \cdot 10^{15}\ \mathrm{s}^{-1}$, and damping constant $3.5 \cdot 10^{14}\ \mathrm{s}^{-1}$ 77
Fig. 4.8 Contributions of electronic, atomic, dipolar, and ionic excitations to the dielectric function 78
Fig. 4.9 Reflection and refraction on the interface between medium 1 with refractive index n_1 and medium 2 with refractive index n_2, with $n_1 < n_2$ 81

Fig. 4.10 Reflection and refraction on the interface between medium 2 with refractive index n_2 and medium 1 with refractive index n_1, with $n_1 < n_2$ 82
Fig. 4.11 Wavelength-dependence of the refractive index of the two optical glasses N-BK7 and SF6 from SCHOTT AG, Germany 84
Fig. 4.12 Reflection and refraction on curved surfaces 85
Fig. 4.13 Chromatic aberration of a standard lens and an achromatic lens, both with focal length $f = 50$ mm 86
Fig. 4.14 Sketch of a grating with partial waves diffracted in the direction given by the diffraction angle β 87
Fig. 4.15 Defining the geometry of the BRDF 89
Fig. 5.1 Principle of a chromatic white light sensor 96
Fig. 5.2 Intensity maximum in the reflectivity measured with a chromatic white light sensor. The solid line represents the fit on the data for finding the peak wavelength position 97
Fig. 5.3 Measuring the thickness of transparent workpieces or transparent coatings with the chromatic white light sensor 99
Fig. 5.4 (**a**) Electronics (light source, spectrometer, display) of a chromatic confocal sensor, (**b**) Chromatic probes with various measuring ranges, (**c**) Multiple point chromatic white light sensor (192 focus points in a line, electronics and optics all inclusive), (**d**) Compact chromatic white light sensor with electronics and optics all inclusive. (Courtesy of Precitec Optronik, Neu-Isenburg, Germany) 100
Fig. 5.5 Areal measurements with the chromatic white light sensor FRT CWL: (**a**) topography of micro prisms, (**b**) topography of a test wafer with various test structures, (**c**) topography of a MEMS microfluidics component (**d**) topography of a Fresnel lens. (Courtesy of FRT GmbH, Bergisch Gladbach, Germany) 101
Fig. 5.6 Areal measurements with the multiple point chromatic white light sensor FRT SLS: (**a**) artificial leather, measurement time 0.2 s, (**b**) wafer with solder balls, measurement time 0.4 s. (Courtesy of FRT GmbH, Bergisch Gladbach, Germany) 102
Fig. 5.7 Areal measurements with two opposite chromatic white light sensors FRT CWL: asphere on top side and segmented free form optics on bottom side of an optical element. (Courtesy of FRT GmbH, Bergisch Gladbach, Germany) 102
Fig. 5.8 Sketch of the principle of an autofocus sensor 103
Fig. 5.9 Detection spot on the PSD, left: out of focus, far distance; mid: in focus, right: out of focus, close distance 103
Fig. 5.10 (**a**) Principle of a point triangulator, (**b**) Point triangulator that fulfills the Scheimpflug condition 104
Fig. 5.11 Principle of a line projection sensor 106

Fig. 5.12 (**a**) Sketch of a modern fringe projection system with two cameras, (**b**) Commercially available system ATOS Triple Scan of GOM GmbH, Braunschweig, Germany. (Picture courtesy of GOM GmbH) 107
Fig. 5.13 Measurement results with the ATOS Triple Scan: (**a**) inspection of a trunk lid, (**b**) inspection of parts of a car body. (Pictures are courtesy of GOM GmbH, Braunschweig, Germany) 108
Fig. 5.14 Optical path in the classical microscope with camera as image detector 109
Fig. 5.15 Abbe's theory of image formation based upon diffraction by the object 110
Fig. 5.16 Illustration of the depth of field in imaging 111
Fig. 5.17 Sketch of a Wollaston prism (left) and a Nomarski prism (right) 113
Fig. 5.18 Illustration of the principle of DIC for transmitted light 114
Fig. 5.19 Optical path in a confocal point sensor 115
Fig. 5.20 Illustration of the PSF for geometrical-optical confocality and wave-optical confocality in a confocal point sensor or a confocal microscope 116
Fig. 5.21 Point-spread function in the conjugated focal plane scanned in steps of Δz of the displacement of the objective 117
Fig. 5.22 Areal measurement with the confocal point sensor FRT CFP on a micro injection moulding. (Courtesy of FRT GmbH, Bergisch Gladbach, Germany) 118
Fig. 5.23 Optical path of a confocal microscope (**a**) with scanning mirrors, (**b**) with rotating multi-pinhole disc (Nipkow disc), (**c**) with rotating array of microlenses 119
Fig. 5.24 Schematic drawing of a Nipkow disk 120
Fig. 5.25 Laser Scanning Confocal Microscope OLS5000. (Courtesy of Olympus Europa SE & Co. KG, Hamburg, Germany) 121
Fig. 5.26 Areal measurements with the confocal microscope FRT CFM: (**a**) part of a microfluidics component with circular bumps (20× objective), (**b**) laser inscription dot on a wafer (100× objective). (Courtesy of FRT GmbH, Bergisch Gladbach, Germany) 122
Fig. 5.27 Principle of focal depth variation 124
Fig. 5.28 (**a**) InfiniteFocus® and (**b**) µCMM utilizing focal depth variation, (**c**) Full form measurement based on Real3D technology of precision rotary component, (**d**) 3D hole measurement on the example of an injection valve. (Pictures are courtesy of Alicona Imaging GmbH, Raaba, Austria) 126
Fig. 5.29 Working principle of a scanning near-field optical microscope (SNOM) 127
Fig. 5.30 Phase shifted interferograms 130

Fig. 5.31 Setups for form inspection with a Twyman-Green interferometer (left) and a Fizeau interferometer (right) 132
Fig. 5.32 Setups for form inspection of spheres, aspheres, and freeforms with a Twyman-Green interferometer (left) and a Fizeau interferometer (right) 133
Fig. 5.33 Graphical representation of the first 21 Zernike polynomials and their meaning 135
Fig. 5.34 Schematic setup of the Tilted-Wave-Interferometer 136
Fig. 5.35 Intensity correlogram at one pixel for a displacement of the objective of 30 μm 137
Fig. 5.36 Sketch of the Mirau-type white light interferometer 140
Fig. 5.37 Sketch of an objective with integrated Michelson interferometer (left) and of a Linnik objective (right) 141
Fig. 5.38 (**a**) Smart WLI for industrial applications, (**b**) 3D topography of an insert (**c**) High resolution roughness measurement on a roughness standard, (**d**) 3D topography of a cutting edge. (Pictures are courtesy of Gesellschaft für Bild- und Signalverarbeitung (GBS) mbH, Ilmenau, Germany) 142
Fig. 5.39 Principal setup of a telecentric white light interferometer with telecentric objective 143
Fig. 5.40 Exemplaric measurements with a parallel beam WLI: (**a**) roughness in an internal cone, (**b**) eveness of a sealing. (Pictures courtesy of ISRA VISION AG, Darmstadt, Germany) 144
Fig. 5.41 Intensity correlogram of a substrate surface coated with a thin transparent film 145
Fig. 5.42 Wavelength scanning interferometer 145
Fig. 5.43 Spectral intensity interferogram at one pixel (n, m) for a point $P(x_n, y_m)$ on the workpiece 146
Fig. 5.44 Principle of increasing the range of unambiguity by using multiple proximate waves (blue and cyan wave) resulting in a beat with strongly enlarged beat wavelength (red wave) 148
Fig. 5.45 (**a**) MWLI sensor, (**b**) MWLI measuring system of AMETEK GmbH – BU Taylor Hobson, Dept. Luphos, Weiterstadt, Germany 149
Fig. 5.46 Example of measurements with a MWLI sensor: surface topography of an aspheric lens with a diameter of 160 mm. (**a**) Photograph of the sensor in action, (**b**) Measured topography scan top view, (**c**) Measured topography scan 3D view. (Courtesy of AMETEK GmbH – BU Taylor Hobson, Dept. Luphos, Weiterstadt, Germany) 149
Fig. 5.47 Prism interferometer for form inspection of planar surfaces 150
Fig. 5.48 Representative examples of measurement with a prism interferometer. Top: component of a hydraulic system, Bottom: single lever faucet ceramics. (Courtesy of LAMTECH Lasermesstechnik GmbH, Stuttgart, Germany) 151

Fig. 5.49 Diffractive grazing-incidence interferometer 152
Fig. 5.50 Dual-sided diffractive grazing-incidence interferometer 153
Fig. 5.51 Principle of frequency scanning interferometry 154
Fig. 5.52 Configuration of reflection DHM (left) and transmission DHM (right) and the corresponding commercial DHM instruments (bottom). (Courtesy of LyncéeTec SA, Lausanne, Switzerland) 156
Fig. 5.53 Examples of DHM measurements with the Reflection DHM® of LyncéeTec SA. Courtesy of LyncéeTec SA, Lausanne, Switzerland 158
Fig. 5.54 Principle of a conoscopic sensor 159
Fig. 5.55 Projection of the wavefront of a plane wave (top) and a spherical wave (bottom) onto a camera chip by a microlens array 160
Fig. 5.56 Optical setup for quality control of lenses using a Shack-Hartmann sensor 161
Fig. 5.57 Measurement example with a Shack Hartmann sensor. (Courtesy of Optocraft GmbH, Erlangen, Germany) 162
Fig. 5.58 Left: Shack-Hartmann wavefront sensor SHSLab. Right: SHSInspect 2Xpass for the measurement of lens systems as microscope lenses, mobile phone lenses, etc. (Courtesy of Optocraft GmbH, Erlangen, Germany) 162
Fig. 5.59 Deflectometric inspection principle. (Courtesy of ISRA VISION AG, Darmstadt, Germany) 164
Fig. 5.60 Deflectometric measurements of (**a**) an exterior car mirror (with profile), (**b**) the flatness of a silicon wafer. (Courtesy of ISRA VISION AG, Darmstadt, Germany) 165
Fig. 5.61 Measuring principle (left) and schematic setup (right) of a Makyoh sensor 166
Fig. 5.62 Makyoh topography image of a plane reference mirror (left) and a deformed Si wafer surface (right) after projection of a regular pattern on the surface 167
Fig. 5.63 Surface roughness characterization based upon scattering 168
Fig. 5.64 Schematic diagram showing regimes of r.m.s. roughness R_q and measurable surface properties that can be derived from light scattering measurements 168
Fig. 5.65 Sketch of the TIS principle in reflection using a Coblentz sphere 169
Fig. 5.66 Sketch of the principle of ARS using a goniometer setup 170
Fig. 5.67 ARS (top) and the corresponding PSD (bottom) for three differently rough surfaces with low, medium, and high roughness 171
Fig. 5.68 Measuring principle (left) and photograph of the ARS sensor OS 500 (right). Pictures are courtesy of OptoSurf GmbH, Ettlingen, Germany 172

Fig. 5.69 Correlation between surface roughness and A_q (top) and between surface inclination and form angle M (bottom). (Courtesy of Optosurf GmbH, Ettlingen, Germany) ... 172
Fig. 5.70 Formation of speckles (left) and typical speckle pattern (right) ... 173
Fig. 5.71 Speckle contrast curves for three different wavelengths ... 174
Fig. 5.72 Cross correlation coefficient in spectral correlation method for four different wavelength differences ... 175
Fig. 5.73 Scheme of the speckle-based optical roughness sensor and scattered light patterns of surfaces with a) $S_a = 3$ nm and b) $S_a = 12.5$ nm. Courtesy of BIMAQ, Bremen, Germany ... 175
Fig. 5.74 (**a**) Sketch of the principle setup for reflectometric measurement with a miniaturized spectrometer, (**b**) Picture of the setup of a MCS 600 spectrometer system. (Courtesy of Carl Zeiss Spectrometry GmbH, Jena, Germany) ... 178
Fig. 5.75 Reflection at a thin layer with refractive index $n_1 + i\kappa_1$ on a substrate with refractive index $n_2 + i\kappa_2$ ... 179
Fig. 5.76 Example for the reflectance of a thin film of silica SiO_2 ($d = 500$ nm and $d = 1000$ nm) on silicon substrate ... 180
Fig. 5.77 Power spectral distribution of the reflectance spectra in Fig. 5.76. The peak at pixel no. 0 is artificially decreased ... 182
Fig. 5.78 Measured reflectance of a photoresist film with $d = 644.2$ nm on glass substrate (left) and the calculated size dependence of the quadratic deviation for this sample (right) ... 184
Fig. 5.79 Measured reflectance spectrum of a 500 nm thick film of SiO_2 on Si (black) and the results of the regression analysis using refractive index data of SiO_2 (blue) and of Si_3N_4 (red) for the film ... 185
Fig. 5.80 Sketch of an ellipsometric thin film measurement ... 189

Fig. 6.1 Schematic setup of an image processing system ... 200
Fig. 6.2 Individual types of illumination: (*1*) coaxial illumination, (*2*) ring light, (*3*) dark field, (*4*) diffuse illumination, (*5*) point light source ... 201
Fig. 6.3 Sketch of the principle of shape from shading ... 204
Fig. 6.4 Defect recognition by shape from shading. (Courtesy of SAC Sirius Advanced Cybernetics GmbH, Karlsruhe, Germany) ... 205
Fig. 6.5 Principle of spectral scanning HSI with hyperspectral cube ... 206
Fig. 6.6 Principle of spatial (pushbroom) scanning HSI with hyperspectral cube ... 207
Fig. 6.7 Principle of a Scanning Electron Microscope (SEM) ... 208
Fig. 6.8 Interaction of the electron beam with the target ... 209

Fig. 6.9 Exemplaric SEM pictures: (**a**) ductile forced fracture in a glass-fiber reinforced plastics, (**b**) shrink hole (bubble) in red bronze, (**c**) stress-corrosion cracking in brass, (**d**) laser burned hole in steel. The pictures in (**a, c**) are courtesy of the Institut für Schadenverhütung und Schadenforschung der öffentlichen Versicherer e.V. (www.ifs-ev.org), Kiel, Germany, and in (**d**) is courtesy of the Laserzentrum der Fachhochschule Münster, Steinfurt, Germany 210
Fig. 6.10 Sketch of the Spectral Domain OCT 212
Fig. 6.11 Sketch of the Swept Source OCT 213
Fig. 6.12 Examples of OCT measurements: (**a**) topography of a microfluidic part, (**b**) all five layers in a transparent multilayer stack, (**c**) topography of a microfluidic channel system in polycarbonate, (**d**) a hole in a transparent film, top view and side view. (The pictures are courtesy of Flo-ir GmbH, Oberdorf, Switzerland) 214
Fig. 6.13 Typical setup of THz-TDS 215

Fig. 7.1 Multiple sensors (chromatic white light, combined confocal microscope and white light interferometer, film thickness sensor) in front of a structured wafer on a x-y stage with aperture. (Courtesy of FRT GmbH, Bergisch Gladbach, Germany) 220
Fig. 7.2 Multisensor capable metrology tool MicroProf® FE for front end and back end production in the semiconductor industry. (Courtesy of FRT GmbH, Bergisch Gladbach, Germany) 222

Fig. 1 Graphical representation of complex numbers 224

List of Tables

Table 1.1 Morphological deviations according to DIN 4760:1982 8

Table 1.2 Cut-off wavelength, evaluation length $L_{n,}$ and tracing (sampling) length L_t in dependence upon the expected R_a or R_z value 24

Table 1.3 Symbols indicating surface lay and its orientation 27

Table 1.4 Human senses and thc rcsult of sensing 32

Table 1.5 Haptic surface sensing and its result 33

Table 1.6 Optical surface sensing and its result 33

Table 2.1 Deviation of the tip shape from the spherical shape measured in grades 44

Table 5.1 Properties an their rating of a chromatic white light sensor 98

Table 5.2 Experimental results for thc numerical-aperture factor f(NA) from Creath [87] 138

Table 5.3 Maximum thickness d_m of an unsupported absorbing layer of different materials for a signal-to-noise ratio $SNR = 1000$. Optical constants of the materials from [194–196] 181

Table 5.4 Results of the thickness determination from regression analysis of measured reflectance spectra in comparison to the values of standards from PTB, Germany 186

Chapter 1
Introduction to Surfaces and Surface Metrology

I often say that when you can measure what you are speaking about, and express it in numbers, you know something about it; but when you cannot measure it, when you cannot express it in numbers, your knowledge is of a meagre and unsatisfactory kind; it may be the beginning of knowledge, but you have scarcely in your thoughts advanced to the state of science, whatever the matter may be.
Lord Kelvin (1824–1907)

Abstract Components and systems used in and manufactured for technical applications consist of two distinct areas, a *bulk* and a *surface*. The surface of a workpiece forms the interface to the surrounding. Its status is not invariant. Rather, water (vapor) and air and all in water or air solved or dispersed substances as well as temperature effects alter the surface gradually even without human influence. Changes that even affect the whole workpiece, like melting, corrosion, or abrasion, start at the surface. A technical surface can also carry signatures of processing techniques like grinding, lapping, or pitch polishing. Moreover, the workpiece can be given new properties by structuring, deposition, or coating that affect primarily the surface. In summary, the surface often determines the functionality and the appearance of the component and is of importance for the proper function and performance of the manufactured component. Therefore, it is essential for product development, process control, quality control, and failure analysis to examine the surface of a workpiece.

Already in the first third of the twentieth century many examinations on technical surfaces were carried out and published. In 1936 Gustav Schmaltz published the book *Technische Oberflächenkunde* [1] where he carefully reviewed all known insights and findings up to this date. With this book he made a big contribution to the development of surface metrology as an autonomous scholarship.

M. Quinten, *A Practical Guide to Surface Metrology*, Springer Series in Measurement Science and Technology,
https://doi.org/10.1007/978-3-030-29454-0_1

For examination of the surface several measuring techniques had been developed. Some of them mainly analyze the chemical composition of the surface. This group of methods includes

- Scanning Electron Microscopy with Energy Dispersive X-ray Analysis (EDX)
- Secondary Ion Mass Spectroscopy (SIMS),
- Glow Discharge Spectroscopy (GDS),
- Auger Electron Spectroscopy (AES),
- Photo Electron Spectroscopy (UPS, XPS, ESCA),
- Small Angle X-ray or Neutron Scattering (SAXS, SANS), and some more.

Another group takes an image of the surface and evaluates the 2D information with means of digital image processing for pattern recognition and defect inspection. For the main part this is done in industrial image processing. A second well-established imaging method for pattern recognition and defect inspection on surfaces is scanning electron microscopy. Further imaging methods well-established in other branches are hyperspectral imaging, optical coherence tomography, and terahertz spectroscopy that become more and more relevant also for surface analysis.

The biggest group of methods is the group of methods for examination of form deviations, topography, roughness, and waviness. It comprises

- Tactile Profiling,
- Scanning Atomic Force Microscopy (AFM),
- Capacitive and Inductive Surface Profiling,
- Confocal Optical Profiling,
- Light Sectional Methods,
- Various Microscopy Methods,
- Various Interferometric Methods,
- Wave Front Sensing,
- Deflectometry, and
- Elastic Light Scattering.

Complementary to these methods, spectral reflectometry and ellipsometry for determination of layer thickness and optical constants are used.

The intention of this book is to introduce in most of these surface metrology techniques except of those for the chemical composition and methods for hardness and friction. But before going into detail of the measuring techniques in the following chapters, this chapter provides first a closer look on the surface itself in a microscopic view and a macroscopic view and on the measurement of surface properties and its validation.

1.1 Microscopic View on a Surface

Considering solid state matter it can be distinguished between a regular arrangement of the atoms which can be described by an elementary cell that is regularly continued within the bulk, and an amorphous structure where elementary cells are not regularly continued but are cross-linked at variable distances. Looking primarily at the regular crystalline structure, the crystal lattice, each atom has a distinct number of next neighbors as long as it is in the bulk. This is dominantly caused by the electronic structure of the single atom and the electronic interaction among the atoms. Each atom is arranged in the crystal so that the resulting electronic states are states of equilibrium of minimal energy.

When approaching the end of the bulk, this is the *surface* as interface between the solid matter and the surrounding, typical properties of a bulk solid exhibit an abrupt change at a surface. One of the most evident changes occur in the crystalline order of the bulk. It gets disturbed since the atoms in the outmost atom layer do have a lower number of neighbors. Then, the electronic interaction among the atoms leads to a new arrangement of the atoms in the surface with electronic states of minimal energy that are different from that in the bulk. Photoelectron spectroscopy actually reveals that the atoms in the surface exhibit so-called *surface states*.

In electrochemistry and nanoscience the *jellium model* [2] is a well-established simple model for separating the conduction electrons of a metal body from the ions. In this model the ions appear as a constant positive charge background with the task to compensate the integrated negative charge of the electrons. While the *ionic* body has an atomically sharp surface, the *electronic* body exhibits a smooth transition of the electron density due to the finite length of the electron waves, resulting in negative charge density *outside* the body. This is called *spill-out effect*. The electron density varies in an oscillating manner, the so-called Friedel oscillations. This is illustrated in Fig. 1.1 where the gray region is the constant positive charge background. As a consequence from the jellium model the surface of a metal body is soft and the surface region cannot longer be approximated by a sharp two-dimensional plane but forms a three-dimensional extended area. Hence, the question may arise what is in general the extension of the surface? Is it only the outmost atom layer or does it extend into the bulk? How deep does it extend into bulk? The answers to these questions are not clear. The extension of the surface namely depends upon the considered property as the above example of the electron density of metals shows.

While the different atomar arrangement in the surface only extends over maximum three or four atomic layers, the relevant surface is quite larger when considering interaction with electromagnetic radiation. E.g., the skin depth in metals is in the order of 20–30 nm for radiation from the soft UV to the radio frequency range. After this distance the intensity of the radiation is diminished to $1/e^2 = 0.135$ of the maximum intensity. Metallic nanoparticles with sizes in this order or metallic films

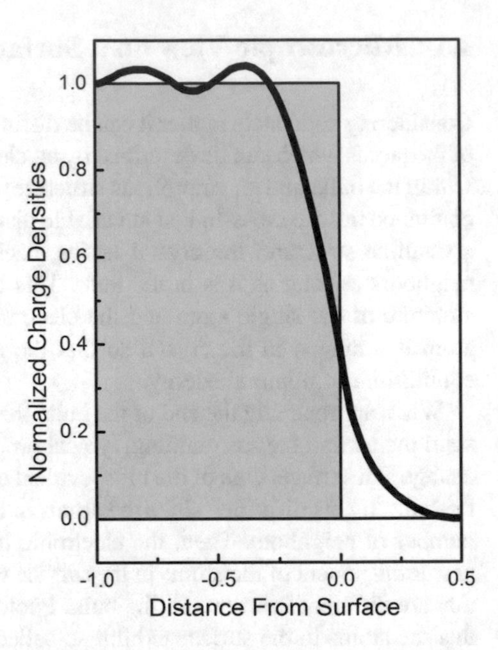

Fig. 1.1 Normalized charge densities of ions and free electrons at a metal surface according to the jellium model

with thickness in this order are therefore "surface", a surface that extends over 100–150 atomic layers! Another extreme example is the recently discovered graphene. Graphene is a monolayer of linked benzene rings with extraordinary properties. For this atomar carbon layer the difference between surface and bulk has vanished in the third dimension. But also for machined surfaces different boundary layers can be distinguished. The inner boundary layer is the zone where the workpiece is altered due to the machining process. It is covered by the outer boundary layer, a very thin zone where the atoms and molecules are exposed to external forces and chemical reactions. Below the inner boundary layer the structure of the workpiece is that of the crystal lattice.

The extent of the surface is relevant also in manufacturing. E.g., when manufacturing white light LEDs from gallium nitride (GaN) the GaN is deposited on synthetic sapphire or spinel. However, the crystalline structures of GaN and sapphire do not match. The mismatch causes enormous stress and strain at the interface of both materials that can lead to detachment of the GaN. Therefore, GaN is deposited in an epitaxial process where the first GaN atoms brought up form atomar layers that gradually change from the crystalline structure of the sapphire to the crystalline structure of GaN. Stress and strain play also an important role when coating a semiconductor wafer surface with a thin film of even a few 10 nm thickness. The enormous stress of several GPa deforms the wafer which on the other hand affects the further processing of the wafer.

The atoms in the outermost surface area of a common workpiece are permanently in contact with the atoms and molecules in the surrounding medium. In consequence, chemical reactions like oxidation start at the surface. Furthermore, atoms and molecules from the surrounding can condensate on the surface, are adsorbed on the surface, and can build thin contamination films. This is important for example in atomic force microscopy. Adsorption processes are involved in almost all technological processes in which surfaces play a crucial role. The most prominent example is heterogeneous catalysis since usually the reactants have to adsorb on the catalyst before they can react. Traditionally, one-dimensional potential curves are used to describe adsorption. The most prominent one goes back to the semi-empirical formula of John Lennard-Jones [3] that describes the interaction potential between atoms with distance R. It is a combination of the attractive potential of the van-der Waals forces ($\propto -1/R^6$) and a repulsive potential ($\propto 1/R^m$, with m mostly $m = 12$).

But not only atoms and molecules from the surrounding play a role. A certain number of atoms of the solid matter leaves the solid matter and forms a gas in front of the solid. The maximum number of atoms or molecules that can leave the solid matter is given by the temperature dependent saturated vapor pressure. It increases with temperature. In thermal equilibrium the atoms in the gas phase permanently condensate on the surface but are replaced by other atoms evaporating from the surface region. Condensation and evaporation are driven not only by the temperature but also by the *surface tension* or better the *surface free energy* that is proportional to the area and forces evaporation, and the energy for building of bulk matter that is proportional to the volume and forces condensation. The interplay of condensation and evaporation is theoretically described by the nucleation theory of Volmer and Flood [4]. Many effects like the Ostwald ripening in colloidal matter or the latent image in the photographic process can be explained by this way.

Another important quantity of a surface is the *work function*. It corresponds to the minimum energy needed to bring an electron from the solid to a point close to the surface but outside the solid surface. The work function is a characteristic property of the surface of the material depending on the crystal face and the contamination of the surface.

Finally, the *surface diffusion* is a general process that involves the motion of adatoms, molecules or atomic clusters at the surface. Similar to the bulk diffusion it is a thermally promoted process with rates increasing with increasing temperature. As the surface diffusion rates and mechanisms are affected by a variety of factors such as the strength of the surface-adparticle bond, the orientation of the surface lattice, attraction and repulsion between surface species, and chemical potential gradients, it is important, e.g. in surface phase formation or epitaxial growth.

1.2 Macroscopic View on a Surface

Unlike for natural surfaces, for technical surfaces manufacturing, processing, and conditioning will leave traces on the surface of a workpiece so that it will deviate from the intended surface. One distinguishes three types of technical surfaces. They are illustrated in Fig. 1.2.

The *nominal surface* is the intended target surface when manufacturing the workpiece. The shape and extent of a nominal surface are usually shown and

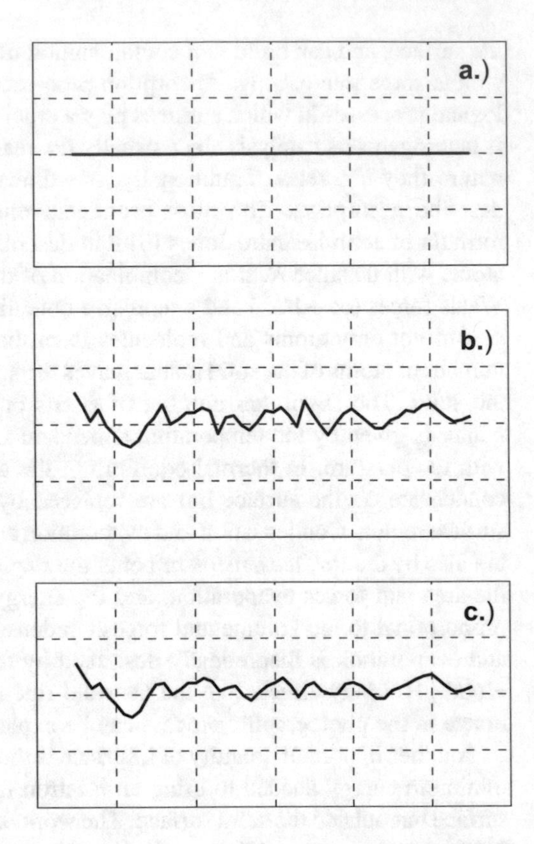

Fig. 1.2 Illustrations of (**a**) the nominal surface, (**b**) the real surface, (**c**) the measured surface

dimensioned on a drawing. The *real surface* is the actual surface obtained after manufacturing the workpiece. The real surface differs from the nominal surface since it carries signatures of the processing. The *measured surface* is a representation of the real surface obtained with some measuring instrument. Each measurement method yields an image that is influenced by the measuring method and its resolution. Hence, different measuring methods may result in different images of the same real surface. Are the obtained surface characteristics then actually representative? Of course, a comparison is then possible if the measurements are carried out with the same measurement parameters and the same evaluation methods.

Beyond the deviations caused by the processing the manufactured workpiece may additionally exhibit deviations in its shape from the ideal shape given in the design drawing. These *shape deviations* can be divided in *coarse shape deviations* and *smooth shape deviations*. Both can further be divided into sections as shown in Fig. 1.3.

Dimensional (Gauge) Deviations are deviations of the dimensions of the workpiece from the tolerances, e.g. the length of a cylinder. *Position Deviations* are

Fig. 1.3 Division of shape deviations

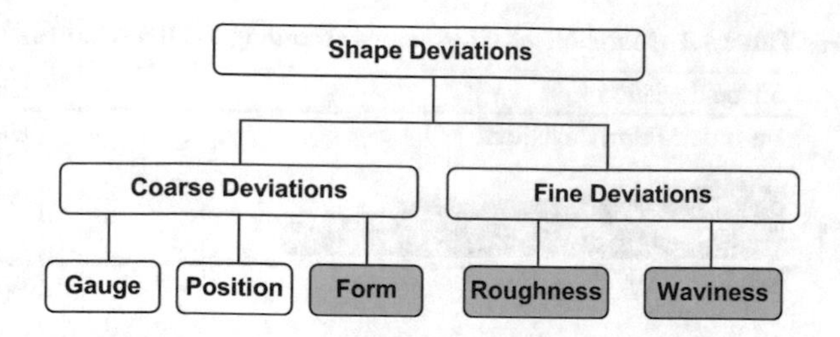

deviations in the position of a single element of the workpiece from the ideal position, e.g. the actual position of drill holes.

Form Deviations are deviations of the real surface from the nominal surface. They include straightness, flatness, roundness, cylindrical form, line shape, and area shape. These deviations result from large scale problems in the manufacturing process such as errors in machine tool ways, guides, or spindles, insecure clamping, inaccurate alignment of a workpiece, or uneven wear in machining equipment. They can affect the performance or the lifetime of a workpiece.

Waviness and Roughness are deviations of the real surface from the nominal surface excluding position and form deviations. Roughness includes the finest (shortest wavelength) irregularities of a surface. Roughness generally results from a particular production process or material condition, e.g. the movement of the cutting tool. Waviness is an unwanted effect coming from the machine tool and is almost always present.

Surface topography measurement is mainly concerned with form deviations, waviness, and roughness. *Topography measurement* means the exact quantitative determination of the geometry and/or the micro structure of technical surfaces. Roughness and waviness characterize the optical or haptic impression of the surface of the specimen. These quantities are compared with norm values that give information on an average behavior of the surface.

According to the German Industry Norm DIN 4760 [5] all shape deviations are classified into six regimes. The morphological deviations up to the fourth order are summarized in Table 1.1. The fifth order deviation (microporous structure) and the sixth order deviation (lattice structure) are not considered in surface metrology but are subject of materials engineering.

Besides the introduced coarse and fine shape deviations a third group of deviations from the nominal surface exist that are not characteristical for the manufacturing process but are important for the effectiveness of the surface: *defects*. Defects comprises scratches, bumps, and dings as well as contamination particles.

Dimensional deviations, position deviations, and defect inspection are commonly investigated by imaging methods with *image processing* as the main process for measurement and evaluation.

Table 1.1 Morphological deviations according to DIN 4760:1982

Shape deviation	Kind of deviation
1st order: Form deviation	Unevenness, ovality, unplanarity
2nd order: Waviness	Waves
3rd order: Roughness (grooves)	Grooves
4th order: Roughness	Cones, corrugations

1.3 Measurement and Validation

The increasing miniaturization, more precise diagnostics, and the increasing level of automation lead on the one hand to more and more properties that must be checked and on the other hand to faster and more complex measuring systems with highly sophisticated software for automated measurement and evaluation. A few examples of measuring tasks and the fields of relevance are listed in the following.

Electronics
Today electronics is based upon semiconductors. Electronic parts are manufactured on semiconductor wafers for which the properties of these wafers are relevant also for the electronic components. Therefore, this industry branch has a lot of measuring tasks that are partly strongly demanding. The tasks comprise measurement of wafer bow, warp, and total thickness variation (TTV), stress, measurement of local step heights, critical dimensions of vias and trenches, or defect inspection and pattern recognition to mention some of the most important.

Illumination
The field of illumination is nowadays also dominated by semiconductors, mainly as light emitting diodes (LED), and by OLEDs (organic light emitting diodes). For both an efficient coupling out of the generated light is of big interest for what also roughness measurement plays a role besides the measuring tasks already mentioned before for electronics.

Energy
The limited availability of some energy resources, the increasing demand on energy production that is harmless for the climate, and the risks of nuclear energy are the impelling forces for the search for alternative energy resources. The most important

yet are solar energy and wind energy. For their effectiveness the measurement and inspection of lengths, step heights, roughness, waviness, defects, and film thickness are important.

Beyond the primary resources and their effectiveness the storage of energy, particularly electrical energy, is of big interest. Here, defect inspection is a main task.

Mobility
Not only for automobiles but also for aircrafts and trains the investigation of manufactured surfaces is of great importance. Mostly it is the reliability of manufactured parts that requires a full and intense inspection since a malfunction of such a part may cause serious injury or even may lead to the death of a human being. But also some sensations (color, haptic impression of the car dashboard, etc.) are of interest. For their effectiveness the measurement and inspection of lengths, step heights, roughness, waviness, defects, and film thickness are important.

Optics
The mode of operation of optical components like lens systems and mirrors strongly depends on the shape, the constitution, and the optical properties of the components. Hence, topography, roughness, waviness, and thickness and performance of coatings are the most prominent tasks.

Medical Engineering
Components manufactured for medical applications are as critical as for the mobility above since a malfunction of such a part may cause serious injury or even may lead to the death of a human being. Therefore, a full and intense inspection of topography, roughness, waviness, defects, and film thickness is important.

In surface metrology exist different measuring techniques with a number of various sensors. Roughly, four relevant groups of methods can be distinguished:

- Methods based upon mechanical interaction,
- Methods based upon electrical and magnetic fields,
- Methods based upon interaction with light or electromagnetic radiation, and
- Imaging methods and image data processing.

Mechanical interaction presumes contact of the probe with the surface. In surface metrology the corresponding methods are the *tactile surface profiling* with a stylus tip and the *atomic force microscope*. The tactile surface profiling plays an outstanding role as it is approved since many decades and is a reliable measuring technique that is the best understood dimensional measurement so that even many common national and international standards in surface metrology developed on the base of measurements with stylus instruments. Also the atomic force microscopy is a distinguished method as it belongs to a few methods with atomic resolution. Both methods are described in Chap. 2.

Methods based upon electrical and magnetic fields are rare but widely used. In fact, mainly sensing with *capacitive probes* and sensors using *eddy currents* have become relevant for surface metrology. Capacitive probes are pretty sensitive on small changes in the distribution of an electric field whereby this method is established in surface metrology. The principles of these methods are described in Chap. 3.

The most methods in surface metrology are based upon interaction of light or other electromagnetic radiation (UV, NIR, IR) with the surface of the workpiece. Several different optical effects are utilized to get a certain optical response from the surface. A part of them record only the intensity reflected or scattered by the surface, another part uses interference of electromagnetic waves to obtain a characteristic response from the surface. Beyond these methods other methods like reflectometry and ellipsometry that are originally not intended to be used in surface metrology, increasingly become important in surface metrology either as stand-alone method or as add-on, due to surface coatings with special functions. As the variety of optical surface metrology methods and its principles is as large it seems appropriate to have first an introduction into the physical basics of optics and optical sensors in Chap. 4 and then to discuss the various optical surface metrology methods in Chap. 5.

Also imaging of the surface with subsequent image data processing is a well-established method to characterize a surface and to determine various kinds of deviation from the ideal (target) surface. Chapter 6 comprises some commonly used and modern methods.

Each method yields a set of data that must be evaluated for the corresponding quantity of interest. With respect to this some statistical quantities are worth to be briefly introduced as they play a role for each kind of measurement on technical surfaces. The statistical quantities that commonly are most important in connection with measurements on technical surfaces are the *resolution*, the *accuracy*, the *repeatability*, and the *reproducibility*.

Resolution means the capability of a measuring system to detect the smallest change of a measured quantity. E.g., the lateral or spatial resolution refers to the question where things are. For a microscope it is approximately given by the mean wavelength λ of the used light source and the numerical aperture NA of the microscope objective as $Res \approx \lambda/NA$. Adjacent parts of the object can only be recognized separately if their sizes or distances are at least in the order of Res.

Resolution may be defined differently. E.g., for the lateral resolution in microscopy there exist three definitions, the *Rayleigh criterion* with $Res = 0.61 \cdot \lambda/NA$, the *Abbe criterion* with $Res = 0.50 \cdot \lambda/NA$, and the *Sparrow criterion* with $Res = 0.47 \cdot \lambda/NA$. Figure 1.4 illustrates the differences between these three criteria for a wavelength of 560 nm and $NA = 0.8$ ($Res = 427$ nm, 350 nm, or 329 nm). According to the Rayleigh criterion the maximum of the second diffraction signal lies in the first minimum of the first diffraction signal, while for the Abbe and Sparrow criterion the diffraction signals lie closer to each other. Hence, only for the Rayleigh criterion the peaks remain well separated.

The *accuracy* of a measuring system is defined over a significant number of repeated measurements on a certified standard. In technical applications this number is commonly quite low with at least 25 repeated measurements. Then, the accuracy is defined as the modulus of the difference between the mean value taken from the repeated measurements and the certified value of the standard as illustrated in

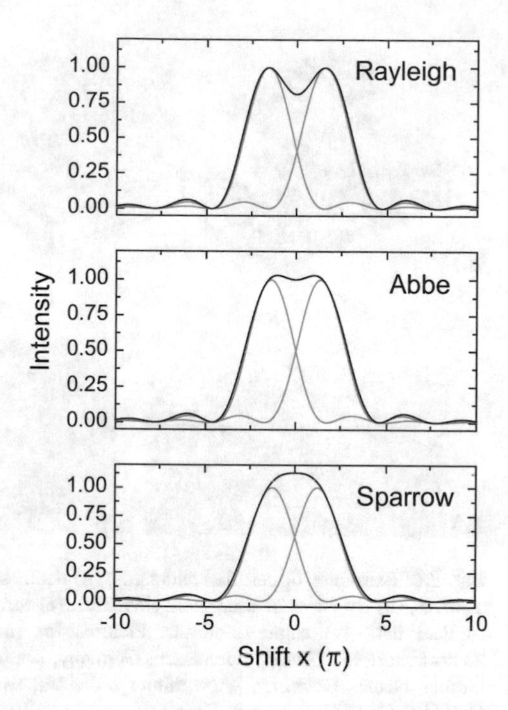

Fig. 1.4 Illustration of the lateral resolution in microscopy according to the Rayleigh criterion, the Abbe criterion, and the Sparrow criterion for a wavelength of $\lambda = 560$ nm and a numerical aperture of $NA = 0.8$

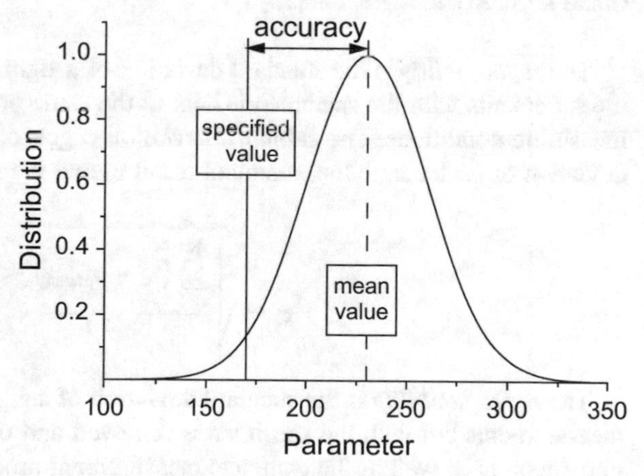

Fig. 1.5 Illustration of the definition of accuracy

Fig. 1.5. E.g., the certified value of a length standard is 1.002 mm and the mean value from the repeated length measurements is 1.035 mm. Then the accuracy of this length measuring system amounts to 0.033 mm.

Certified standards are available for heights/depths (grooves standard), roughness, planarity (optical flat), parallelism and thickness, and angle. Examples are shown in Fig. 1.6. For their use at customer's site they are measured in a laboratory accredited as calibration laboratory by a national gauging institute (e.g. PTB in Germany, NIST in the U.S.A., METAS in Switzerland, etc.).

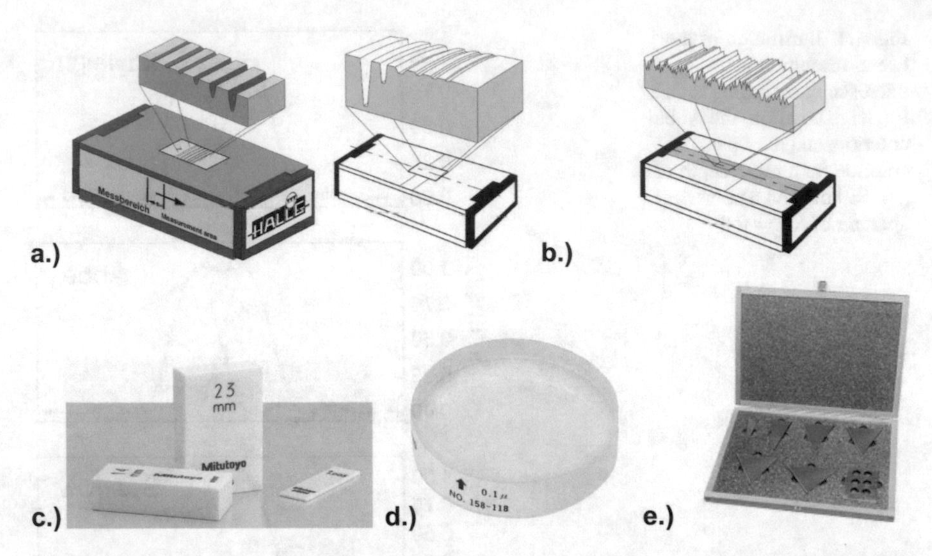

Fig. 1.6 Examples of certified standards: (**a**) depth setting standards with round and rectangular grooves, (**b**) rougness measurement standards, (**c**) ceramic end-gauge blocks, (**d**) flatness standard (optical flat), (**e**) angle standards. Pictures for (**a** and **b**) courtesy of HALLE Präzisions-Kalibriernormale GmbH, Edemissen, Germany, for (**c** and **d**) courtesy of Mitutoyo Deutschland GmbH, Neuss, Germany, www.mitutoyo.de, and for (**e**) courtesy of SARTORIUS Werkzeuge GmbH & Co. KG, Ratingen, Germany

The *repeatability* is the standard deviation of a significant number *N* of repeated measurements with the specimen is kept at the same position and with unmodified measuring conditions. The *standard deviation* σ gets calculated from the quadratic deviation of each single measurement result z_n and the mean z_{mean}:

$$\sigma = \sqrt{\frac{\sum_{n=1}^{N} (z_n - z_{mean})^2}{N-1}} \tag{1.1}$$

The *reproducibility* is the standard deviation of a significant number of repeated measurements but with the specimen is removed and newly placed for each single measurement as well as the complete measurement process is repeated. In that case also the mean value of the repeated measurements may be different to that obtained in repeatability measurements. As a rule of thumb it is $\sigma_{reproducibility} \approx 3 \cdot \sigma_{repeatability}$.

Besides these most important statistical quantities further quantities are in use and relevant. These are the *uncertainty*, the *stability*, and the *tolerance*.

There exist different definitions of the *uncertainty* [6–8]. The uncertainty *U* is an estimated domain within the true value lies with a given probability. Known systematic errors are excluded from the estimation since they can be removed or corrected. Uncertainty of measurement comprises in general many components.

The first contribution to the uncertainty is the type A standard uncertainty u_A. It bases on the statistical analysis of a series of observations and is connected with the repeatability, i.e. with an observed frequency distribution

$$u_A = \frac{\sigma}{\sqrt{N}} \tag{1.2}$$

It corresponds to the standard deviation of the mean value z_{mean}.

The second contribution, the type B standard uncertainty u_B, is determined from non-statistical influences as the measurement setup, the environment, the specimen, the user, and the measurement strategy. Each of these n components can also be assigned a standard deviation u_n. This standard deviation can be estimated from assumed probability distributions (mostly rectangular distribution) and from experience or other information. The pool of information includes

- Previous measurement data,
- Experience with or general knowledge of the behavior and properties of relevant materials and instruments,
- Manufacturer's specifications,
- Data provided in calibration and other certificates, and
- Uncertainties assigned to reference data taken from handbooks.

The type B standard uncertainty u_B follows from the estimated uncertainties as

$$u_B^2 = \sum_n u_n^2 \tag{1.3}$$

According to the Gaussian law of error propagation the *combined standard uncertainty* u_c is obtained from all type A and type B uncertainties to

$$u_c = \sqrt{\sum_n \left(\frac{\partial f}{\partial X_n}\right)^2 \cdot u_n^2} \tag{1.4}$$

This presumes that the measurand Z can be modeled as function f of input parameters X_n. Otherwise, if the partial derivatives cannot be calculated, each partial derivative is replaced by the value 1.

Finally, the *expanded uncertainty* U is given by the combined standard uncertainty u_c weighted with a *coverage factor* k

$$U = k \cdot u_c \tag{1.5}$$

The coverage factor k is a measure for the probability. For $k = 1$ the probability is 68.3%, for $k = 2$ it is 95%, for $k = 3$ it is 99%. Mostly $k = 2$ is used.

The result of a measurement is then conveniently expressed as $Z = z_{mean} \pm U$, which is interpreted to mean that the best estimate of the value attributable to the measurand Z is z_{mean}, and that $z_{mean} - U$ to $z_{mean} + U$ is an interval that may be expected to encompass a large fraction of the distribution of values that could reasonably be attributed to Z.

Stability is the change of the accuracy caused by altering conditions, e.g. wear, temperature, air pressure, relative humidity, etc. over a time period.

Tolerance is the limit for the accepted accidental deviation of the measured value from the nominal value. For example, when manufacturing a specimen with nominal length of 200 mm the manufacturing tolerance T of ± 0.2 mm means that the specimen (not the measurement on the specimen!) need not be less than 199.8 mm or larger than 200.2 mm. The compliance of tolerances is required in order that components can be combined without adaptation.

When measuring a characteristics of the workpiece the accuracy of the measurement must always be small compared to the manufacturing tolerance. There are two rules of thumb: accuracy $\leq$ a tenth of the manufacturing tolerance and resolution $\leq$ a tenth of the accuracy.

Finally, in practice it is important that the measurements are traceable. *Traceability* ensures that measurements are consistent and accurate. Any quality system in manufacturing will require that all measurements are traceable and that there is documented evidence of this traceability. According to the ISO/IEC Guide 99:2007–12 [7] traceability is the property of the result of a measurement whereby it can be related to stated references, usually national or international standards, through a documented unbroken chain of comparisons all having stated uncertainties. As all the measurements in surface topography measurement are related to length, the documented unbroken chain of comparisons begin with a standard with stated uncertainty for a length certified by a national institute for standardization or an accredited laboratory. For the traceability of surface topography measuring instruments two parts must be considered: the traceability of the instrument and the traceability of the analysis algorithms and parameter calculations. Instrument traceability is achieved by calibrating the axes of operation of the instrument using certified standards or calibration artifacts. Traceability for parameter calculations can be carried out using calibrated artifacts having associated parameters.

To rate a measuring equipment often a test code for determination of accuracy and repeatability under consideration of the manufacturing tolerance T is applied. There exist three rating methods.

In method 1 two values are determined, the *potential gage capability index* C_g and the *critical gage capability index* C_{gk}. C_g is a measure for the repeatability and C_{gk} a measure for the accuracy, both under consideration of the manufacturing tolerance. They are defined as

$$C_g = \frac{0.2 \cdot T}{6 \cdot \sigma} \tag{1.6}$$

$$C_{gk} = \frac{0.1 \cdot T - |z_{mean} - z_0|}{3 \cdot \sigma} \tag{1.7}$$

where z_{mean} is the mean of a significant number of repeated measurements and z_O is the nominal value of the used certified standard. $|z_{mean} - z_O|$ is the accuracy. The definitions in Eqs. (1.6) and (1.7) may be slightly modified depending on the internal rules in the factories. The measuring equipment is suited if the accuracy and the repeatability fulfill certain conditions. The minimum condition is given by the requirement that $C_g = C_{gk} = 1$. In practice it is even requested that $C_g \geq 1.33$ and $C_{gk} \geq 1.33$, meaning that $\sigma \leq 0.025 \cdot T$ for C_g and $0.25 \cdot |z_{mean} - z_O| + \sigma \leq 0.025 \cdot T$ for C_{gk}.

In method 2, the so-called GR&R = Gage Repeatability & Reproducibility study, the repeatability, reproducibility, and the overall margin of error are determined manually (not automated) at which user influences are considered. Some users carry out multiple tests with multiple repetition parts with the same measuring equipment. The intention is to use $n \geq 10$ repetition parts, $k \geq 3$ operators, and $r \geq 2$ repetitions. The GR&R study determines the *precision* P as geometrical average of the standard deviations of operators and repetitions. The value gets related to the manufacturing tolerance T

$$GRR\% = \frac{P}{T} \tag{1.8}$$

The measuring equipment is suited if $GRR\%$ is less than 10%, and is acceptable if $GRR\%$ remains less than 30%.

Method 3 is also a GR&R study where user influences are not considered. It is used in automated measurements where the position of the workpiece remains unaffected. The intention is to carry out automated measurements with $n \geq 25$ repetition parts and $r \geq 2$ repetitions.

As already shown in Fig. 1.2 the measured surface differs from the real surface as it is the image of the real surface after measurement. Each measurement method yields an image that is influenced by the measuring method and its resolution. To discuss the main alterations caused by the measurement standardization of the measuring methods plays an important role for comparability of measured quantities. International, national, and even industrial norms have been developed to establish a certain standard and a certain degree of comparability.

Most prominent norms for surface metrology are developed from the International Standardization Organization [9], abbreviated ISO, Technical Committee 213 (TC 213), dealing with Dimensional and Geometrical Product Specifications and Verification. ISO is an independent non-governmental international organization with a membership of 162 national standards institutes with its headquarters in Geneva, Switzerland. Most of the norms in the wide range of standards for surface texture profile measurements and areal measurements are adopted as European Standards (EN) or by national institutes, e.g. the German Institute for Standardization (DIN). Other relevant norms are from the American Society for Testing and

Materials (ASTM), the American Society of Mechanical Engineers (ASME), SEMI, the global industry association serving the manufacturing supply chain for the electronics industry, or JIS, the Japanese Industrial Standards.

As the determination of the fine structure of the complete surface was not possible in the past one managed with profile measurements on random samples although profiles are directional and do not provide all information. The increasing miniaturization, more precise diagnostics, and the increasing level of automation requests for more detailed and more exact information on the surface as only can be obtained with areal measurements. In the following advantages and disadvantages of both, profile measurements and areal measurements are discussed.

1.3.1 Profile Measurement

The profile measurement has a long tradition in surface metrology so that many common national and international standards in surface metrology developed on the base of profiles as well as many surface characteristics are defined on the base of profiles. The following list summarizes the most relevant profile specification standards from ISO [9] as they stand at the time of writing of this book:

ISO 1302:2002 Geometrical Product Specifications (GPS) – Indication of surface texture in technical product documentation.

ISO 3274:1996 Geometrical Product Specifications (GPS) – Surface texture: Profile method – Nominal characteristics of contact (stylus) instruments.

ISO 4287:2000 Geometrical Product Specifications (GPS) – Surface texture: Profile method – Terms, definitions and surface texture parameters.

ISO 4288:1996 Geometrical Product Specifications (GPS) – Surface texture: Profile method – Rules and procedures for the assessment of surface texture.

ISO 5436 Geometrical Product Specifications (GPS) – Surface texture: Profile method; Measurement standards.

- ISO 5436-1:2000 Part 1: Material measures
- ISO 5436-2:2012 Part 2: Software measurement standards

ISO 12085:1996 Geometrical Product Specifications (GPS) – Surface texture: Profile method – Motif parameters.

ISO 12179:2000 Geometrical Product Specifications (GPS) – Surface texture: Profile method – Calibration of contact (stylus) instruments.

ISO 13565 Geometrical Product Specifications (GPS) – Surface texture: Profile method; Surfaces having stratified functional properties.

- ISO 13565-1:1996 Part 1: Filtering and general measurement conditions
- ISO 13565-2:1996 Part 2: Height characterization using the linear material ratio curve
- ISO 13565-3:1998 Part 3: Height characterization using the material probability curve

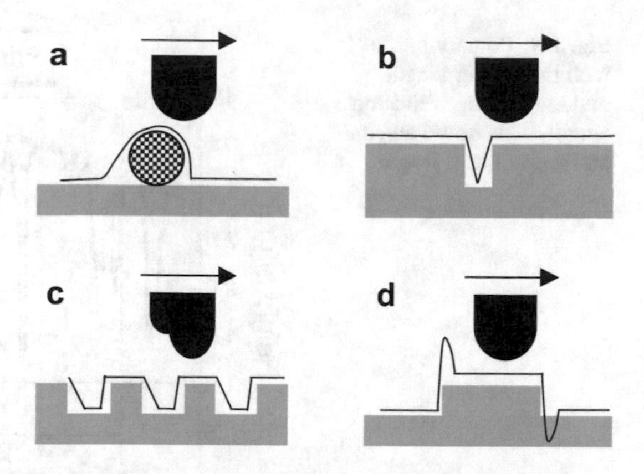

Fig. 1.7 Effects that influence the measured profile. For description see text

ISO 16610 Geometrical product specifications (GPS) – Filtration.

ISO 16610-21:2012 Part 21: Linear profile filters: Gaussian filters
ISO 16610-22:2012 Part 22: Linear profile filters: Spline filters
ISO 16610-28:2010 Part 28: Profile filters: End effects
ISO 16610-31:2009 Part 31: Robust profile filters: Gaussian regression filters
ISO 16610-32:2009 Part 32: Robust profile filters: Spline filters
ISO 16610-61:2012 Part 61: Linear areal filters: Gaussian filters
ISO 16610-71:2011 Part 62: Robust areal filters: Gaussian regression filters

It is meaningful to give here a brief overview on the alterations of the real surface when measuring a profile, and how the evaluation of profile measurements works according to standardized evaluation methods.

When measuring a profile either the probe (tip, light spot, etc.) or the sample is moved along a certain line of distinct length and measure the variation of the detected height along this line (height distribution). The recorded profile may be affected by different effects that originate from either surface structures, contaminations, or even the measuring system. In Fig. 1.7 some effects and their influence on the measured profile are summarized.

As the motion of the probe is unidirectional contaminations or surface structures with round shapes may become asymmetrically reproduced (Fig. 1.7a). The lateral extension of the probe may be larger than structures in or on the surface. Then, the number of significant data points is reduced such as these structures are not or incompletely reproduced (Fig. 1.7b). Form deviations of the probe will be reflected in the profile particularly at the edges of surface structures (Fig. 1.7c). At sharp edges of surface structures the recorded signal may overshoot and/or undershoot leading to "bat wings" (Fig. 1.7d).

The first step in the evaluation chain of the recorded profile is to remove the influence of such systematic errors if they are known. The profile just measured and corrected for systematic influences is further corrected with respect to the form of the workpiece and further alterations that originate from the measurement setup. The

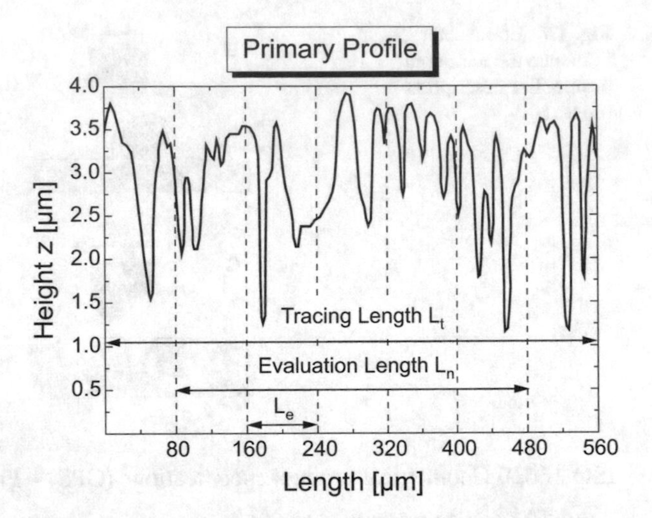

Fig. 1.8 Primary profile with the tracing length divided into an evaluation length, a pre travel length, and a post travel length

most prominent is an inclination of the workpiece for what the profile is also inclined. Inclinations can be removed by fitting a straight line on the profile and subtract this line from the data points. The form of the workpiece is optionally removed from the data by a standardized form filter with parameter λ_f. The form filter is a low pass filter and suppresses all structures larger than the parameter λ_f. Further manipulations of the recorded data can be carried out using various filters (despike, median, smoothing, FFT, etc.), but note that each filter will have influence on the forthcoming parameters to be determined like roughness or waviness parameters.

The profile that is now obtained is called *primary profile*. It is shown in Fig. 1.8. It represents the height distribution along the sampling line without further manipulation. The profile has the *tracing length* or *sampling length* L_t. From this profile already surface characteristics, the *primary parameters*, can be derived. For this the next step is to divide the tracing length L_t into three parts: a pre travel length L_1, an *evaluation length* L_n, and a post travel length L_2.The evaluation length L_n is further divided into five identical parts L_e. Commonly, $L_1 = L_2 = L_e = L_t/7$. Therefore, it is recommended to select a tracing length as a multiple of 7. The pre travel length and the post travel length are optional. L_1 and L_2 take into account possible transient and decay processes when starting and finishing the measurement.

The separation of roughness and waviness is then carried out using standardized filters applied on the evaluation length L_n. One distinguishes two filters: a noise filter and a roughness filter. The noise filter separates the noise in the measurement from the profile. Hence, it is a high pass filter that discriminates all structures smaller than a certain size given by the parameter λ_s. All larger structures pass this filter. The application of a noise filter is optional. The main filter is the roughness filter that separates roughness from waviness. For this a parameter λ_c, called *cut-off*

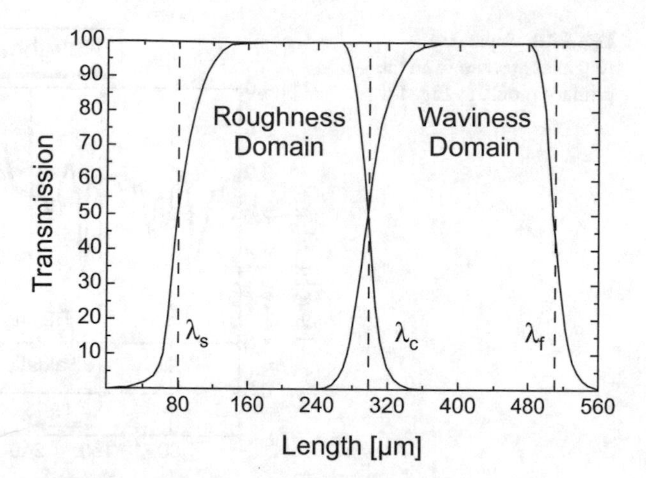

Fig. 1.9 Transfer characteristic and mode of operation of the noise filter, roughness filter, and form filter

wavelength, must be given. All structures in the profile smaller than λ_c get evaluated as roughness and all larger structures as waviness. By default λ_c is chosen as $\lambda_c = L_e$ but can be chosen arbitrarily.

The standardized roughness filter is a low pass filter with a weighting function or *transfer characteristic s(x)* according to ISO 16610-21 (ISO 16610 replaces ISO 11562) that discriminates all structures with length $x > \lambda_c$:

$$s(x) = \frac{1}{\alpha\lambda_c} \cdot \exp\left(-\pi \cdot \left(\frac{x}{\alpha\lambda_c}\right)^2\right) \tag{1.9}$$

with $\alpha = \sqrt{\frac{\ln 2}{\pi}}$. The corresponding transfer characteristic for the waviness is $1 - s(x)$. The actual filtering is carried out in the Fourier space with the Fourier transform of *s(x)*

$$S(\lambda) = \exp\left(-\pi \cdot \left(\frac{\alpha\lambda_c}{\lambda}\right)^2\right) \tag{1.10}$$

The transfer characteristic and its Fourier transform for the form filter and noise filter are similarly defined with the corresponding parameters λ_f and λ_s. The filtered data in the Fourier space are finally transformed back into real space. The mode of operation of noise filter, roughness filter and form filter in the real space is illustrated in Fig. 1.9.

The separated roughness and waviness profiles from the above primary profile are shown in Figs. 1.10 and 1.11.

After having separated the roughness from the waviness some surface characteristics can be calculated in the interval given by the evaluation length L_n. In the

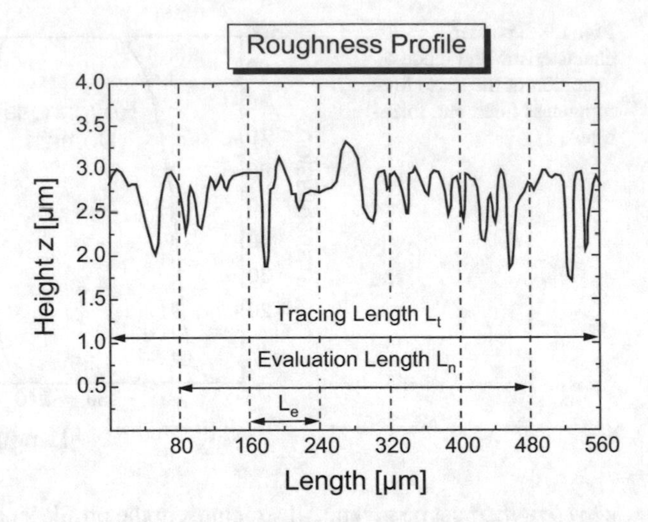

Fig. 1.10 Separated roughness profile from the primary profile in Fig. 1.8

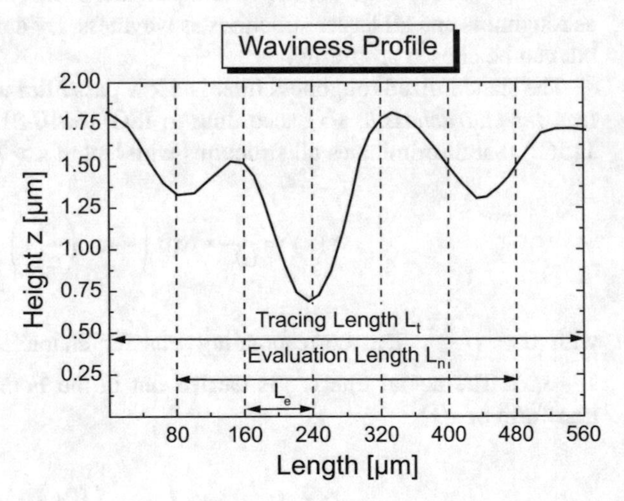

Fig. 1.11 Separated waviness profile from the primary profile in Fig. 1.8

following, definitions of some selected relevant quantities for the roughness (R) according to the standards ISO 4287 and ISO 4288 are given. The corresponding waviness (W) and primary (P) quantities are similarly defined. The parameters are obtained for an exemplaric roughness profile shown in Fig. 1.12 together with its height distribution histogram. The exemplaric roughness profile is chosen so that the mean value (the middle line) is zero.

Arithmetic Mean Deviation R_a (W_a, P_a)

The *arithmetic mean deviation* R_a is the average of all deviations of the roughness profile from the middle line along the sampling line. For this the modulus of $z(x) - z_{mean}$ is calculated for each position x. Then, R_a is obtained as

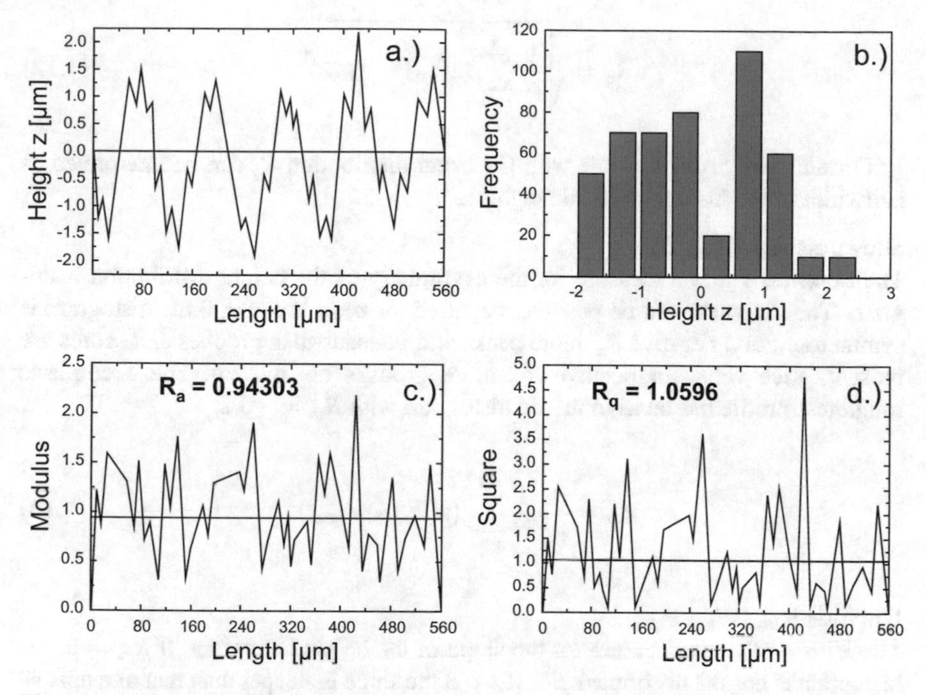

Fig. 1.12 Illustrations of (**a**) an exemplaric roughness profile, (**b**) its height distribution histogram, (**c**) the parameter R_a, (**d**) the parameter R_q

$$R_a = \frac{1}{N}\sum_{n=1}^{N}|z(x_n) - z_{mean}| \tag{1.11}$$

It is also known as Arithmetic Average (AA) or Center Line Average (CLA). R_a does not tell the whole story about a surface. Often different surfaces that behave very differently in some applications may have the same R_a. Therefore, to distinguish between surfaces that differ in shape or spacing other parameters must be calculated for a surface that measure peaks and valleys and profile shape and spacing. The more complicated the shape of the surface is the more sophisticated other measuring parameters must be considered.

Root Mean Square Deviation R_q (W_q, P_q)

The *quadratic mean deviation* R_q or r.m.s. (root mean square) is the average of all quadratic deviations of the roughness profile from the middle line along the sampling line. For this the square of $z(x) - z_{mean}$ is calculated for each position x. Then, R_q is obtained as

$$R_q = \sqrt{\frac{1}{N}\sum_{n=1}^{N}(z(x_n) - z_{mean})^2} \tag{1.12}$$

For an ideally rough profile with Gaussian distribution R_q can be interpreted as halfwidth of the histogram of all heights z.

Skewness R_{sk} (W_{sk}, P_{sk})
The *skewness* R_{sk} is a measure for the asymmetry of the height distribution histogram. The skewness can be positive, negative, or zero. If $R_{sk} = 0$ the histogram is symmetric. For a negative R_{sk} more peaks and plateaus than grooves and scores are present. Vice versa for positive R_{sk} more grooves are present. The exemplaric roughness profile has an asymmetric histogram with $R_{sk} = -0.2$.

$$R_{sk} = \frac{1}{R_q^3}\frac{1}{N}\sum_{n=1}^{N}(z(x_n) - z_{mean})^3 \tag{1.13}$$

Kurtosis R_{ku} (W_{ku}, P_{ku})
The *kurtosis* R_{ku} is a measure for the slope of the height histogram. If $R_{ku} = 3$, the histogram is normal distributed. For $R_{ku} > 3$ the slope is steeper than that of a normal distribution, for $R_{ku} < 3$ the slope is wider than that of a normal distribution. The exemplaric roughness profile has an asymmetric histogram with $R_{ku} = 1.84$.

$$R_{ku} = \frac{1}{R_q^4}\frac{1}{N}\sum_{n=1}^{N}(z(x_n) - z_{mean})^4 \tag{1.14}$$

Further relevant quantities are given below and illustrated in Fig. 1.13 and Fig. 1.14. They are derived from the exemplaric roughness profile shown in Fig. 1.12.

Maximum Profile Valley Depth R_v (W_v, P_v)
This is the distance between the middle line and the deepest valley.

Maximum Profile Peak Height R_p (W_p, P_p)
This is the distance between the middle line and the highest peak.

Total Height of Profile R_t (W_t, P_t)
This is the distance between the highest peak and the deepest valley. Hence

$$R_t = R_p + R_v \tag{1.15}$$

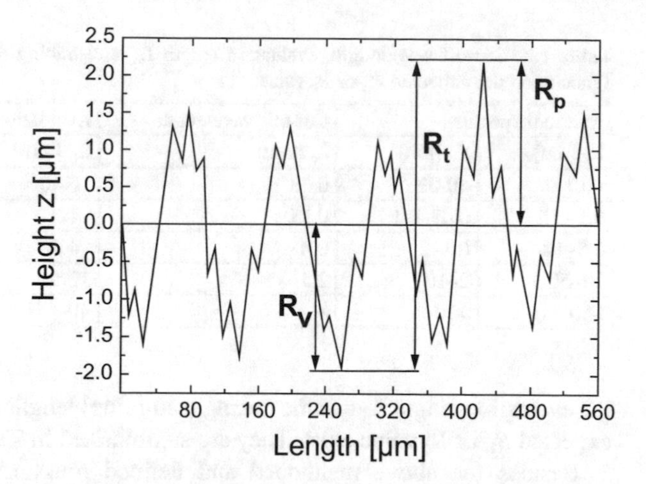

Fig. 1.13 Illustration of the parameters R_v, R_p, and R_t

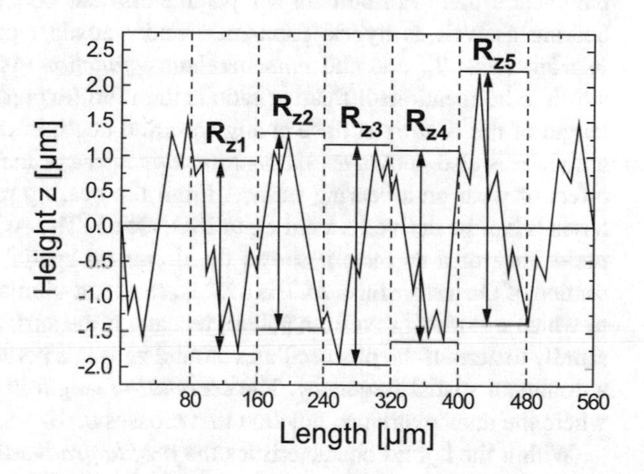

Fig. 1.14 Illustration of how to determine the parameter R_z

Average Peak to Valley Height R_z (W_z, P_z)

For determination of R_z the evaluation length is divided into five parts. For each part the peak to valley height $R_z(n)$ (corresponding to R_t in this part) is determined. Then, R_z is the average of the five $R_z(n)$ values

$$R_z = \frac{1}{5}\sum_{n=1}^{5} R_z(n) \tag{1.16}$$

The maximum of the five $R_z(n)$ values is also a characteristics – the **Maximum Peak to Valley Height R_{max}**.

The above values depend upon the choice of the cut-off wavelength λ_c since all structures in the profile smaller than λ_c get evaluated as roughness and all larger structures as waviness. By default λ_c is chosen as $\lambda_c = L_e$ but can be chosen arbitrarily. For aperiodic profiles recommendations for the cut-off wavelength and

Table 1.2 Cut-off wavelength, evaluation length L_n, and tracing (sampling) length L_t in dependence upon the expected R_a or R_z value

Aperiodic profiles		Cut-off wavelength	Evaluation length	Tracing length
R_z [μm]	R_a [μm]	λ_c [mm]	L_n [mm]	L_t [mm]
<0.1	<0.02	0.08	0.40	0.56
0.1–0.5	0.02–0.1	0.25	1.25	1.75
0.5–10	0.1–2	0.8	4.0	5.6
10–50	2–10	2.5	12.5	17.5
>50	>10	8	40	56

the evaluation length L_n and the tracing (sampling) length L_t in dependence upon the expected R_a or R_z value exist. They are summarized in Table 1.2.

Besides the above mentioned and defined roughness and waviness surface parameters there are still further parameters that consider flatness, spacing, and bearing as well as hybrid parameters and statistical parameters. Particularly the *bearing ratio* T_p and the *autocorrelation function ACF* or *autocovariance* are worth to be mentioned. Bearing ratio is the ratio (expressed as a percentage) of the length of the bearing surface at any specified depth in the profile to the evaluation length. It is also known as *Abbott-Firestone curve* or material ratio. It simulates the effect of wear on a bearing surface. From the bearing ratio further surface characteristics can be derived according to ISO 13565. The ACF is used to determine the periodicity of a surface; it shows the dominant spatial frequencies along a cross section of the test surface. ACF is a measure of self-similarity of a profile – the extent to which a surface waveform pattern repeats. If the surface is random the plot drops rapidly to zero. If the plot oscillates around zero in a periodic manner the surface has a dominant spatial frequency. The *correlation length* is the length along the x-axis where the autocovariance function first crosses zero.

Within the hybrid characteristics the *profile gradients* and *horizontal roughness characteristics* play a certain role particularly when using scattering techniques for measurement of the surface texture. According to ISO 4287 they are defined as.

Average Absolute Slope Δ_a

$$\Delta_a = \frac{1}{N}\sum_{n=1}^{N}\left|\frac{\partial z(x_n)}{\partial x_n}\right| \tag{1.17}$$

Root Mean Square Slope Δ_q

$$\Delta_q = \sqrt{\frac{1}{N}\sum_{n=1}^{N}\left(\frac{\partial z(x_n)}{\partial x_n}\right)^2} \tag{1.18}$$

Average Wavelength λ_a

$$\lambda_a = \frac{2\pi R_a}{\Delta_a} \tag{1.19}$$

Root Mean Square Average Wavelength λ_q

$$\lambda_q = \frac{2\pi R_q}{\Delta_q} \tag{1.20}$$

Correlation Length Λ_k

$$\Lambda_k \approx \frac{\lambda_q}{\sqrt{2\pi}} \tag{1.21}$$

Surface texture is indicated in technical product documentation such as drawings, specifications, contracts, and reports by means of graphical symbols and textual indications. The basic graphical symbol illustrated in Fig. 1.15 should not be used without the inclusion of complementary information on requirements on the surface concerning

- Profile characteristics according to ISO 4287, or
- Motif characteristics according to ISO 12085, or
- Characteristics derived from the bearing curve (ISO 13565-02, −03).

Generally, any manufacturing process is permitted. If material shall not be removed or material removal is even required this is indicated by changing the blue indicated region accordingly as illustrated in Fig. 1.15. Complementary information is written at the positions *a* – *e* with the following meaning.

Position a – one single surface texture requirement

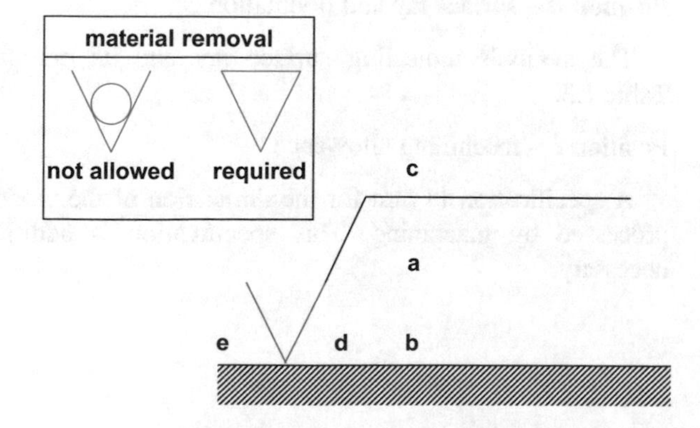

Fig. 1.15 Indication of surface texture according to ISO 1302

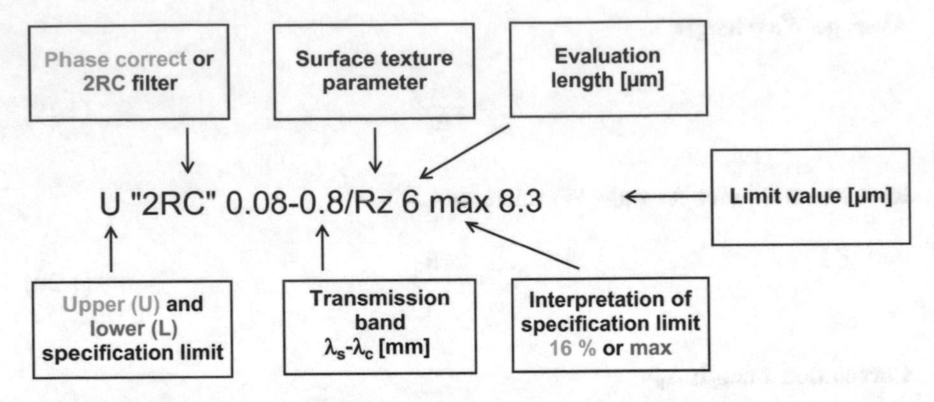

Fig. 1.16 Control elements of surface texture requirement

A surface texture requirement is built up of several different control elements which can be part of the indication expressed on the drawing or the surface texture information given in other documents. The elements are shown in Fig. 1.16. With specification of the cut-off wavelength automatically the standard evaluation length is fixed according to Table 1.2 but the value given in the requirement can deviate from the standard value. The requirements written in green letters are default and have not to be given. The requirements written in blue letters are given if necessary.

The *16%* rule for the specification limit means that up to 16% of the measured values are allowed to exceed or to fall below the given limit. The *max* rule means that no measured value needs to exceed or to fall below the given limit.

Position b – second surface texture requirement

Analogous to Position a. The second surface texture requirement is additional and is indicated if necessary.

Position c – manufacturing method

Specification of the manufacturing process, treatment, or coating. For example: turned, ground, or milled.

Position d – surface lay and orientation

The symbols indicating surface lay and its orientation are summarized in Table 1.3.

Position e – machining allowance

A specification in mm for the dimension of the workpiece if it is yet or again processed by machining. This specification is additional and is indicated if necessary.

Table 1.3 Symbols indicating surface lay and its orientation

Graphic symbol for lay	Description
=	Lay is parallel to the projection plane of the surface to which the symbol is applied
⊥	Lay is perpendicular to the projection plane of the surface to which the symbol is applied
×	Lay is crossed in two oblique directions relative to the to the projection plane of the surface to which the symbol is applied
M	Lay is multidirectional in the projection plane of the surface to which the symbol is applied
C	Lay is approximately circular relative to the center of the surface to which the symbol applies
R	Lay is approximately radial relative to the center of the surface to which the symbol applies
P	Lay is particulate, non-directional, or protuberant

1.3.2 Areal Measurement

Although a profile measurement gives some functional information about a surface, a two-dimensional (2D) or areal measurement of the surface has some benefits compared to profile measurement:

- More repeatable results are obtained,
- More complex parameters are possible,
- More significant parameters are obtained (volume parameters, fractal dimension, autocorrelation and related parameters (S_{al}, S_{tr}, local homogeneity, gradient distribution),
- Less user influence on the results, and finally
- A true 3D image of the surface is obtained with the help of which the manufacturer gains a better visual record of the overall structure of the surface.

Area analysis has also disadvantages:

- Longer measuring time,
- Only small areas may be examined in one shot, and
- Higher computational effort to get the data.

The first attempt to areal measurements is to conduct a profile measurement of a length of N measuring points along the x-axis repeatingly M times along the y-axis with a stepwidth Δy between each profile. This areal measurement can be achieved with each stylus tip, with AFM, and with each optical point sensor as long as either the sensor is moved to the N x M surface points or the sample gets positioned to these points. The advantage of this procedure is that large surface areas can be measured, however eventually on the costs of a long measuring time.

The corresponding roughness and waviness surface area parameters are defined as sR_m, sW_m or sP_m according to the standard ISO 4287, with $m = a, q, sk, su, p, v, t$ etc. As for form deviations and waviness and roughness the absolute position of the surface in space is irrelevant, possible inclinations can be considered by subtracting a best-fit plane from 3D data. For determination of roughness and waviness parameters the next step is again a filtering to separate roughness and waviness using a filter according to ISO 16610. This is done by applying the norm filter first in x-direction and then in y-direction. The calculation of the surface parameters includes a double sum over both directions, e.g. for sR_a

$$\mathrm{sR_a} = \frac{1}{\mathrm{N} \cdot \mathrm{M}} \sum_{\mathrm{m}=1}^{\mathrm{M}} \sum_{\mathrm{n}=1}^{\mathrm{N}} \left| \mathrm{z}(\mathrm{x_n}, \mathrm{y_m}) - \mathrm{z_{mean}} \right| \tag{1.22}$$

A second approach is to use field of view sensors that automatically observe a certain surface area and image the field of view on a detection area, the image field. The best-known representative of this class of sensors is the microscope. Many of the modern field of view sensors are based on the microscope although different techniques are used to get an image of the surface area. Surface characteristics are defined in the norm ISO 25178 Geometrical Product Specification (GPS) – Surface texture: Areal [9]. All areal standards are part of ISO 25178, which consists of at least the following parts:

- Part 1, 2011: Indication of surface texture
- Part 2, 2011: Terms, definitions and surface texture parameters
- Part 3, 2011: Specification operators
- Part 6, 2010: Classification of methods for measuring surface texture
- Part 70, 2011: Material measures
- Part 71, 2011: Software measurement standards
- Part 72, 2011: XML softgauge file format
- Part 601, 2010: Nominal characteristics of contact (stylus) instruments
- Part 602, 2010: Nominal characteristics of non-contact confocal chromatic probe instruments
- Part 603, 2011: Nominal characteristics of non-contact phase shifting interferometric microscopy instruments
- Part 604, 2011: Nominal characteristics of non-contact coherence scanning interferometry instruments
- Part 605, 2011: Nominal characteristics of non-contact point autofocusing instruments
- Part 607, 2011: Nominal characteristics of non-contact focus variation instruments
- Part 608, 2011: Nominal characteristics of non-contact imaging confocal microscope instruments

- Part 700, 2011: Calibration of non-contact instruments
- Part 701, 2010: Calibration and measurement standards for contact (stylus) instruments

The amplitude and height characteristics S_m with $m = a, q, sk, su, p, v, t$ etc. are partly resumed from 2D-profile standards and transferred to areal evaluation. Further characteristics are function related characteristics (height distribution and material ratio curve as well as material ratio parameters), volume characteristics, spatial characteristics, and hybrid characteristics. They are also partly resumed from 2D-profile characteristics and transferred to areal evaluation or are completely new, reflecting evaluation algorithms that cannot be applied to 2D-profiles.

The chain of evaluation is similar to that described before for profile measurement. However, the used nomenclature has changed in which now two kinds of surfaces are distinguished

- The *S-F surface* and
- The *S-L surface*.

The difference lies in the sequence of applying filters.

The new vocabulary contains

1D (profile)	2D (areal)
Roughness–/waviness filter λ_c	S-filter (S for small)
	L-filter
Form filter λ_f	F-filter (F for form) or F-operator
Cut-off wavelength	Nesting index
Evaluation length	Evaluation region
Sampling or tracing length	Definition region
Filtered data	Scale bounded surface

By default, the *definition region* and the *evaluation region* are quadratic with their size given by the *nesting index* of the used filter. Again, after correction for systematic errors the form of the workpiece is (optionally) removed from the 2D data by applying the *F-filter* or *F-operator*. This filter uses various numerical filters that are described in ISO 16610 to separate form deviations of different order by best fit. The value of the F-operator nesting index is typically chosen to be five times the scale of the coarsest structure of interest. The nesting index is a value from the following series: ...; 0.1 mm; 0.2 mm; 0.25 mm; 0.5 mm; 0.8 mm; 1.0 mm; 2.0 mm; 2.5 mm; 5.0 mm; 8.0 mm; 10 mm; ...

The next step in the evaluation chain of areal measurements is to apply a noise filter with λ_s similar to the profile evaluation. This noise filter is optional. The main filter to derive surface roughness and waviness parameters is the *S-filter*. By default a Gaussian filter as for the profile evaluation is used. The value of the S-filter nesting index is typically chosen to be five times the scale of the coarsest structure of interest. The nesting index is a value from the following series: ...; 0.1 μm; 0.2 μm; 0.25 μm;

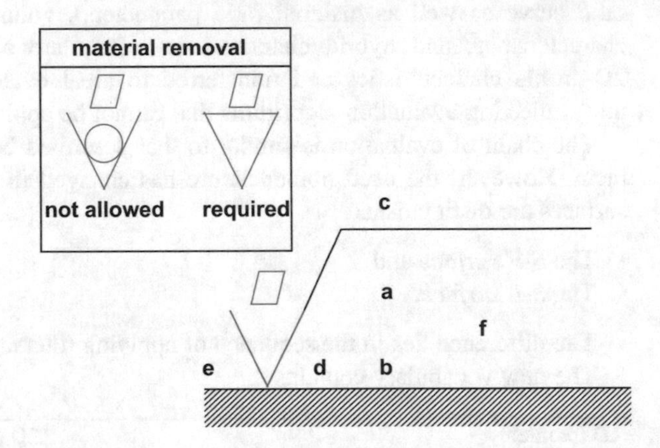

Fig. 1.17 Definition of the S-F surface and the S-L surface according to ISO 25178-2

Fig. 1.18 Indication of surface texture according to ISO 25178-1

0.8 μm; 1.0 μm; 2.0 μm; 2.5 μm; 8.0 μm; 10 μm; . . . The application of F-filter and S-filter defines the *S-F surface*.

The *L-filter* is an additional filter that removes long-wavelength components The value of the L-filter nesting index is typically chosen to be five times the scale of the coarsest structure of interest. The nesting index is a value from the following series: . . .; 0.1 mm; 0.2 mm; 0.25 mm; 0.8 mm; 1.0 mm; 2.0 mm; 2.5 mm; 8.0 mm; 10 mm;. . .

If no form deviations are to be removed the application of the S-filter and optionally the L-filter defines the *S-L surface*.

Filtering is always carried out in two successive steps: first the filter is applied in one lateral direction and then the filter is applied on the filtered data in the perpendicular lateral direction.

Figure 1.17 shows the definition of the S-F surface and the S-L surface according to ISO 25178-02.

A third approach is based on projection of patterns and the evaluation of the distortions of these patterns caused by the surface variations compared to a reference surface. The measuring methods comprised in this class of areal measurement stand out by the area size and the time needed to measure this area. Unless the microscopic methods before, the surface area that can be measured with relatively high accuracy and resolution can be pretty large.

Also for areal measurements the requirements on the surface are indicated in technical product documentation such as drawings, specifications, contracts, and reports by means of graphical symbols and textual indications. According to ISO 25178-01 the basic graphical symbol is modified to distinguish it from that in profile measurements. The new symbol is illustrated in Fig. 1.18.

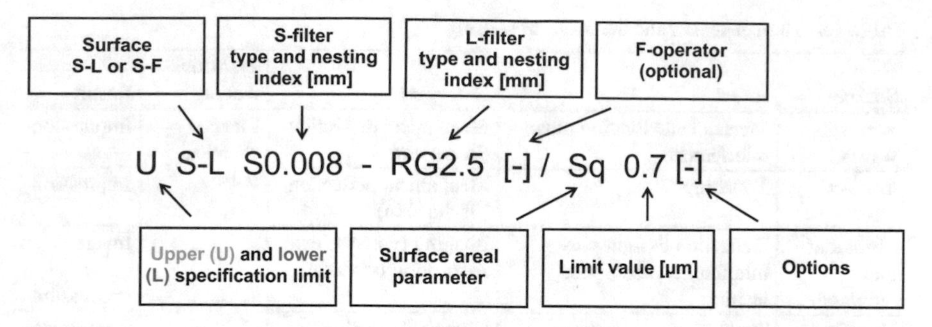

Fig. 1.19 Control elements of areal surface texture requirement

Compared to ISO 1302 a rhombus in the symbol indicates the use for areal measurements. Position a has a new description. Positions b – e are as described before. An additional Position f is introduced.

Position a – one single surface texture requirement

An areal surface texture requirement is again built up of several different control elements. The elements are shown in Fig. 1.19.

The specification of the type of the S-filter means the use of a spline filter (S) according to ISO 16610. The specification of a (default) Gaussian filter (G) is not needed. The specification of the L-filter means a robust Gaussian filter (RG) according to ISO 16610. The nesting index of the L-filter also defines the evaluation area (here 2.5 × 2.5 mm. The surface areal parameter S_q need not to exceed 0.7 µm.

Position f – Projection planes for indication of orientation of the evaluation area and the surface lay

Usually the projection plane for indication of the surface lay coincides with the projection plane of the evaluation area. When indicating areal surface texture it may be possible that one has to define two different projection planes.

1.3.3 Measurement Modules

Up to now terms and definitions of the evaluation of measured data have been considered. Now, modules commonly used for the measurement will be considered.

The human being has five senses to collect information about his surrounding. Each of them gets excited by a source and cause a certain change compared to a previous state. This change is detected and transformed in any case into an electrochemical pulse which generates either an impression or an image. Besides the pure impression (e.g. loud, silent, hard, soft, hot, cold, sour, sweet, delightful) also a direction to the source can be retrieved. The corresponding processes are illustrated in Table 1.4.

Table 1.4 Human senses and the result of sensing

Source	Effect	Detection	Transformation	Result
Acoustic waves	Mechanical vibration plus interference	Mechanical deflection (in the ear)	Electro-chemical pulse	Impression
Contact force	Touching	Mechanical deflection (of the skin)		Impression
Light thermal radiation	Deflection of light plus interference, absorption heating	Imaging (with the eye) absorption (via skin)		Image color impression
Chemical compounds	Tasting	Chemical reaction		Impression
Gases	Smelling	Chemical reaction		Impression

In surface metrology tasting and smelling are not useful. Although sensing of thermal radiation does only yield an impression, the classification in different temperature ranges can be used technologically to get an image of the 2D and even 3D temperature distribution. Acoustic waves and their deflection play a negligible role in surface metrology and are not further considered here.

What can be best reproduced in surface metrology is the haptic sensing (with stylus tip or AFM) and the optical sensing and optical imaging. Both deliver three-dimensional information and an idea of the object under consideration in a virtual 2D or 3D image. In contrast to the human being who is insensitive to electrical or magnetic fields several techniques have been developed that use minute changes in electrical or magnetic fields to record also a virtual 2D or 3D image of a surface.

In the following Tables 1.5 and 1.6 the schemes for haptic sensing and optical sensing are given analogous to Table 1.4.

Obviously, the reproduction of haptic sensing and optical sensing is a complex task that needs corresponding mechanical, electrical, optical, and electronic components. Moreover, a movement either of the sensor or the sample must be established. The most important components and their influence on the measurement are discussed in the following.

1.3.3.1 Stages

In surface metrology often x-y-stages are used to bring the sample to a certain position (x, y) instead of moving the sensor to this position. In any case the movement contributes to the noise in the measurement. This measurement noise limits the capability of the instrument to measure high spatial frequencies on the surface of a sample. Hence, it should be separated from the deviations caused by the surface under consideration. For this an optical flat with a deviation from flatness of less than 30 nm ($\lambda/40$) is required having a certain surface roughness. There are two methods to separate the measurement noise. The first method is based on subtraction [10]. It requires two measurements at the same position on the

Table 1.5 Haptic surface sensing and its result

Haptic sensing	
Source	Mechanical force on surface
Effect	Mechanical deflection due to repulsion or attraction
Detection	Change of electrical fields (piezoelectric transducer)
	Induced electrical current (inductive transducer)
	Deflection of light spot on photosensitive material (analog signal)
Transformation	Discrete voltage steps (digital signal)
Result	Height information (mainly)

Table 1.6 Optical surface sensing and its result

Optical sensing	
Source	Electromagnetic wave (light, UV-, NIR-radiation)
Effect	Deflection by
	Reflection
	Refraction
	Diffraction
	Scattering
Detection	Photoeffect in photosensitive material
	Electrical charges (released electrons) (analog signal)
Transformation	Discrete voltage steps (digital signal)
Result	Height information, thickness information, color information

sample. The two topography data sets are subtracted from each other such that the form and the underlying roughness of the optical flat are eliminated. The measurement noise $S_{q,noise}$ is estimated using

$$S_{q,\,noise} = \frac{S_q}{\sqrt{2}} \tag{1.23}$$

with S_q being the measured r.m.s. roughness of the optical flat.

The second method is based upon the assumption that the noise contribution decreases when averaging multiple measurements of the surface topography at the same location on a sample [11]. The measured S_q of the optical flat is a function of the instrument noise $S_{q,noise}$ and the roughness of the flat $S_{q,flat}$:

$$S_q = \sqrt{S_{q,\,flat}^2 + S_{q,\,noise}^2} \tag{1.24}$$

When carrying out N repeated measurements at the same location on the surface of the flat, the contribution of the instrument noise to the root mean square is decreased by a factor $1/N$ while the contribution of the optical flat is preserved:

$$S_{q,N} = \sqrt{S_{q,\text{flat}}^2 + \frac{1}{N} S_{q,\text{noise}}^2} \tag{1.25}$$

The instrument noise can now be extracted from Eqs. (1.24) and (1.25)

$$S_{q,\text{noise}} = \sqrt{\frac{N}{N-1}\left(S_q^2 - S_{q,N}^2\right)} \tag{1.26}$$

1.3.3.2 Piezoelectrical and Inductive Transducers

The detection of the mechanical deflection of a stylus tip is often based on an piezoelectrical transducer or an inductive transducer producing either a voltage or a current, and an electronic amplifier to boost the signal from the transducer to a useful level. In this case noise generated by the random thermal motion of charge carriers (usually electrons) is unavoidable. Beyond this thermal noise other types of noise can be generated by the electrical or electronical noise and different processes. The most prominent are shot noise, 1/*f* noise (flicker noise), and burst noise. Shot noise always occur if charge carriers must traverse a potential barrier. The reason is that each individual carrier must traverse the gap. This process however is statistically random. It is missing in conductors and resistors. Flicker noise or 1/*f* noise is characterized by its steady decrease with increasing frequency. In electronic devices like MOSFETs it plays a distinct role as it dominates over the thermal noise at frequencies below $f = 15$ kHz. In contrast, burst noise consists of sudden step-like transitions between two or more discrete voltage or current levels at random and unpredictable times. Piezoelectrical and inductive transducers used in surface metrology are mainly affected by shot noise and 1/*f* noise.

1.3.3.3 Optical Detectors

Mechanical deflection of a stylus tip can also be detected using an optical transducer, i.e. the vertical movement of the tip moves an optical element like a small mirror and a laser beam gets deflected. The deflected light is then recorded with an optical detector. In the optical line and areal profilers there is no mechanical deflection but reflection, refraction, diffraction, or scattering brings a certain amount of light to the optical detector.

Optical detectors use the inner photoeffect that releases electrons in a quantity that is proportional to the energy of the photons collected during the exposure time and over the detector area. The released electrons get further amplified and transformed in discrete voltage steps so that optical detectors are also affected by shot noise and 1/*f* noise.

For detection at a single wavelength or a single frequency there are two possible detectors: photomultiplier tubes and electronic detectors based on semiconductor photodiodes.

Photomultiplier tubes are extremely sensitive detectors for light in the ultraviolet, visible, and near-infrared range. They use the principle of the inner photoeffect in a cathode. The released electrons get accelerated in an electric field and hit further electrodes, the dynodes. With these dynodes the electrical current produced by the incident light gets amplified by a factor of up to 10^8 in multiple dynode stages. High gain, low noise, high frequency response or, equivalently, ultra-fast response, and large area of collection are the advantages of photomultiplier tubes. Correspondingly, photomultipliers are essential in low light level spectroscopy, confocal microscopy, Raman spectroscopy, fluorescence spectroscopy, nuclear and particle physics, astronomy, medical diagnostics, and medical imaging.

Semiconductor photodiodes use the principle of the inner photoeffect in a MOSFET (Metal Oxide Semiconductor Field Effect Transistor) with silicon (Si) as semiconductor where electrons get released by light but remain in the material. They change from the valence band into the conduction band and get stored in the potential well, a region in the semiconductor component from which they are readout. For measurements at longer wavelengths in the near infrared the semiconductor indium gallium arsenide InGaAs has been proven to be well-suited. The properties of the released electrons depend on various parameters: the absorption coefficient, the recombination time of the generated electron-hole pairs, the diffusion path, and the chemical and physical structure of the material above the photosensitive layer. Moreover, temperature plays a certain role because electrons in the valence band can also be excited into the conduction band by temperature. This holds true particularly for so-called indirect semiconductors as for example silicon. The temperature released electrons also exist in the case of no illumination. If they get read out the corresponding current is called *dark current* which forms a constant signal level that cannot be used for evaluation. Active cooling of the electronic device clearly diminishes the dark current, which is essential mostly for InGaAs detectors. A further contribution to the dark current signal I_{dark} comes from the downward electronics in the AD-converter. To remove this contribution an initial dark current measurement should be carried out before other calibrations and before measuring any data. The dark measurement should be conducted once the electronics has reached its operating temperature, usually after a few minutes. Software algorithms capture the dark level and store it in memory so that it can be subtracted during measurements. As the dark current level increases with temperature (from ambient and particularly from the internal components of the measuring device) it is recommended to repeat the dark current measurement from time to time.

When illuminating the pixels of the camera a certain number of electrons get released and stored in each pixel. Their number is preferably proportional to the amount H of light striking the photodiode. This amount H is given by the product of illuminance and exposure time. So far, the behavior of the electronic component is quite similar to the photographic film. The corresponding response curve is illustrated in Fig. 1.20.

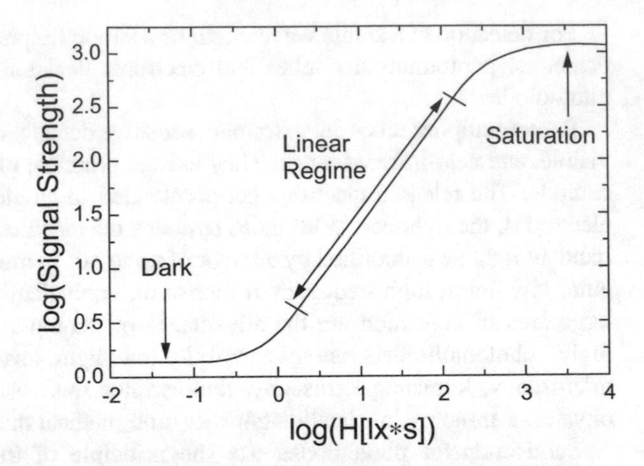

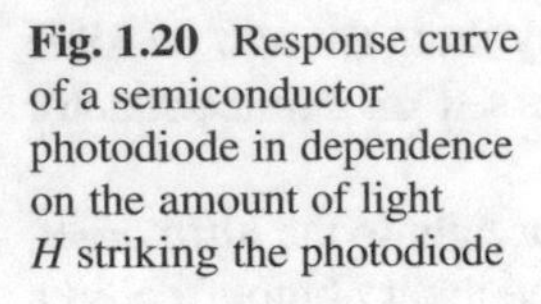
Fig. 1.20 Response curve of a semiconductor photodiode in dependence on the amount of light *H* striking the photodiode

The number of released electrons cannot exceed a maximum level which is given by the capacitance of the pixel. Above the dark current level and below the saturation level a region of nonlinear response restricts the region of linear response.

Digital cameras consist either of a line or an array of diodes called pixel basing on two technologies, CMOS (Complementary Metal Oxide Semiconductor) or CCD (Charged Coupled Device). For both, each pixel works as described before. The difference lies in the charge-to-voltage conversion. For CMOS the conversion already happens in the pixel and the charges have not to be shifted. CCD technique is used to shift the photoinduced charges in many small steps (vertical and horizontal shift registers) to a central analog-to-digital converter. The principle of this shifting is comparable to a chain of buckets: one filled bucket is drained into the next empty bucket in the chain, and so forth. It may happen that the capacitance of a pixel is exceeded and excess charges flow into the pools of neighboring photodiodes. This effect is called *blooming*. Blooming occurs for long-during intensive illumination of the pixel. It can be reduced by additional vertical and lateral overflow drains that direct the excess charges away from the pixel. The disadvantage is that these additional electric lines restrict the size of the pixels and hence the photosensitivity of the pixels.

In Fig. 1.21 the quantum efficiency of a CMOS image sensor is shown. The array has a broad spectral response from 200 nm to approximately 1000 nm wavelength with a high level of ultraviolet sensitivity. The quantum efficiency tends to zero for wavelengths longer than 1000 nm because for the intrinsic photoeffect in Si at least an energy of 1.1 eV (bandgap of Si) is necessary to transport an electron from the valence band to the conduction band. This energy corresponds to a wavelength of 1127 nm.

The quantum efficiency of a CCD sensor is plotted in Fig. 1.22 for a front-illuminated CCD image sensor, a back-thinned CCD image sensor, and a back-thinned linear CCD detector.

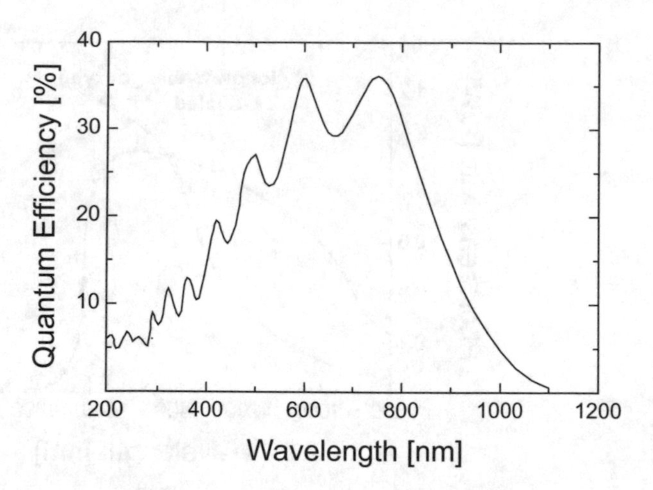

Fig. 1.21 Quantum efficiency of a silicon CMOS image detector in dependence on the wavelength of incident radiation. The oscillations are caused by a thin transparent protective quartz window in front of the detector

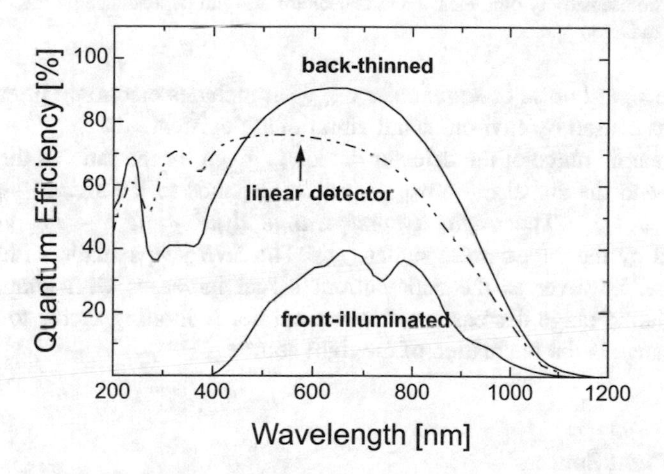

Fig. 1.22 Quantum efficiency of a front-illuminated CCD image sensor, a back-thinned CCD image sensor, and a back-thinned CCD linear detector in dependence on the wavelength of incident light

As can be recognized, the quantum efficiency increases by almost a factor 2 if the photodiode gets thinned at its rear side so that it becomes transparent and then coated by an anti-reflective coating. This design is however more expensive.

The photosensitivity of a linear InGaAs detector array is plotted in Fig. 1.23 versus the wavelength. Depending on the cooling of this sensor the spectral range can be extended from 1700 to 2200 nm but then with less sensitivity at wavelengths between 800 and 1200 nm.

To obtain a measure for the accuracy of a signal value it is common practice to measure repeatingly N times the signal value at a fixed wavelength or frequency with the same constant exposure time. The standard deviation of these N measurements

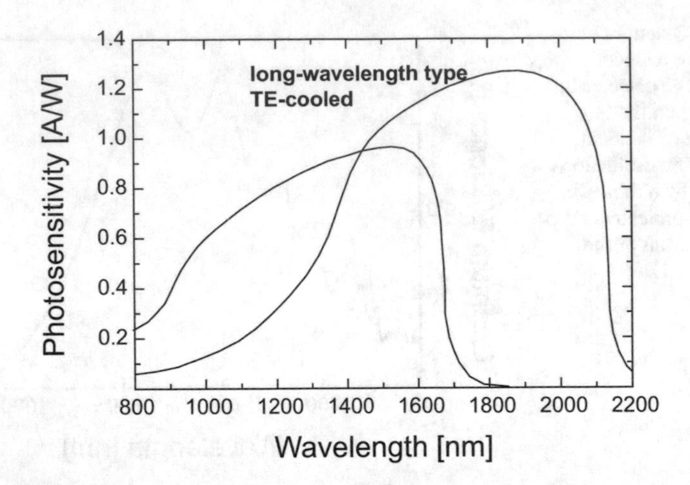

Fig. 1.23 Photosensitivity of a linear InGaAs detector array in dependence on the wavelength of the incident radiation

defines the *signal noise* or *static noise* σ_{Noise}. It includes electronic jitter as well as fluctuations caused by environmental vibrations if existent.

The dynamic range of the detector device is given by the ratio of the saturation signal I_{sat} and the signal noise σ_{Noise} and corresponds to the *signal-to-noise ratio* $SNR = I_{sat}/\sigma_{Noise}$. The *useful dynamic range* $dynR = (I_{sat} - I_{dark})/\sigma_{Noise}$ gets diminished by the dark current signal I_{dark}. The *SNR* keeps stable with increasing temperature, however as the dark current signal increases with temperature the useful dynamic range decreases with temperature. A limiting factor for the useful dynamic range is the fluctuation of the light source.

1.3.3.4 Light Sources

To obtain a most realistic image of the surface the illumination plays an important role. Available light sources are halogen lamps, light emitting diodes (LED) and superluminescence diodes (SLD), (high pressure) arc lamps, deuterium lamps for the near UV, and lasers. Among them lasers and light emitting diodes are the preferred light sources in surface metrology. The artificial light source laser emits strongly coherent monochromatic light which is favorable for methods utilizing interference for generating characteristic patterns from the examined surface. But not in all cases a coherent light source is favorable. Light sources with low coherence are used in white light interferometry to increase the region of unambiguity in the measurement. These light sources are light emitting diodes (LED) having a central emission wavelength and a certain bandwidth. As an example the emission spectra of a white light LED, a cyan LED and two SLDs in the near infrared spectral region are shown in Fig. 1.24. White light LEDs are based on short-wavelength LEDs that exhibit strong emission in the blue visible spectral region being covered with a layer

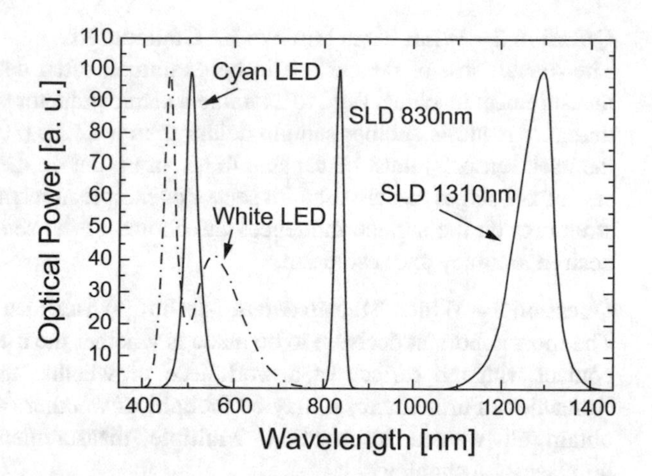

Fig. 1.24 Typical spectra of a white light LED, a cyan LED (center wavelength 520 nm) and two superluminescence diodes with their center wavelength in the near infrared at 830 nm and 1310 nm

of photoluminescent material ("phosphor") that extends the emission to longer wavelengths by phosphorescence. The spectrum in Fig. 1.24 is the typical spectrum of such a white light LED covered with the phosphor YAG:Ce. Outside the corresponding luminescence regions no further radiation is emitted by the LED. The main advantages of LEDs are high luminance, easy to couple into a fiber, and an extremely long lifetime of typically 25,000 h and more.

1.3.4 The Way to Reliable Surface Data

Independently of whether 1D- (profile) measurements or 2D- (areal) measurements are planned a good practice to obtain reliable surface data is to consider first the following questions.

Question 1 – What Shall Be Measured?

- Type of surface:
  polished, grinded, rough, smooth, corrugated, coated, highly reflective, matte, stained, etc.
- Surface geometry:
  flat, round, cylindrical, aspheric, structured, bumps, etc.
- Surface material:
  metal, semiconductor, glass, plastics, transparent, organic, etc.
- Measurement task:
  topography, roughness, waviness form deviations, positions and shifts, layer thickness

The answer to this question is essential for the selection of the measuring equipment. In many cases even multiple different sensors are requested.

Question 2 – What Sizes Have to Be Considered?
The overall size of the surface to be examined often determines the setup of the measurement machine. E.g., to examine a semiconductor wafer of 6 in. (150 mm) in diameter requests another sample holder than a 12 in. (300 mm) wafer. Moreover, the used sensor(s) must be capable to hit any point on the sample. This defines the size of x-y-stages that are used for sensor or sample movement. In the end, the size of structures on the surface influences the choice of the sensor, independently of the desired accuracy and resolution.

Question 3 – Which Measurement Method(s) Shall Be Used?
The most important decision to be made is whether the used method may come into contact with the surface of a workpiece or whether the used method must be contactless. Furthermore, it may be thought of whether one method is sufficient to obtain all wanted information. Multiple measurement tasks usually require multisensor technology.

Question 4 – How Accurate Must Be the Result?
This question is strongly related to Question 3 since any measurement method can only deliver a restricted picture of the surface of a workpiece. The restrictions may be caused by the measuring range of a method or sensor and the measuring rate. Yet, more important is to know the achievable accuracy, resolution, repeatability, and reproducibility of the selected method or sensor.

Question 5 – How Fast the Measurement Shall Be Carried Out?
This question is important particularly for automated measurements. Areal measurements with field of view sensors are fast but have usually small field of views. Stitching of pictures is time-consuming. On the other hand, profiling with a profilometer or a point sensor in the x-y plane can yield information on large areas but is rather slow. Moreover, the sample preparation time is important, for example when using scanning electron microscopy.

Question 6 – Where Shall Be Measured?
To achieve in a certain accuracy and resolution the environmental conditions must be suitable. This concerns the room where the measurement is carried out, either laboratory, production hall, or cleanroom, as well as temperature, relative humidity, vibrations, external electrical and magnetic fields etc. during the measurement. A further aspect is whether the measurement shall be carried out inline or offline.

Question 7 – Who Shall Measure?
Also the answer to this question is relevant for the significance of the measurement results. Trained and skilled personnel is necessary to operate the measuring device or to prepare and to implement measuring recipes for fully automated measurements.

References

1. Schmaltz, G.: Technische Oberflächenkunde – Feingestalt und Eigenschaften von Grenzflächen Technischer Körper Insbesondere der Maschinenteile. Springer, Berlin (1936). This book is part of the digitalization project Springer Book Archives of Springer, Berlin
2. Lang, N.D., Kohn, W.: Theory of metal surfaces: charge density and surface energy. Phys. Rev. B. **1**, 4555–4568 (1970)
3. Lennard-Jones, J.: Processes of adsorption and diffusion on solid surfaces. Trans. Faraday Soc. **28**, 333–359 (1932)
4. Volmer, M., Flood, H.: Tröpfchenbildung in Dämpfen. Z. Phys. Chem. A. **170**, 273–285 (1934)
5. DIN 4760: Gestaltabweichungen; Begriffe, Ordnungssystem (Form Deviations; Concepts; Classification System). Beuth-Verlag, Berlin (1982)
6. ISO Guide 98-3: Guide to the Expression of Uncertainty in Measurement (GUM), International Organization for Standardization (ISO). Central Secretariat, Geneva (1993)
7. ISO/IEC Guide 99:2007-12, International Vocabulary of Metrology – Basic and General Concepts and Associated Terms (VIM), Beuth Verlag GmbH, (2007)
8. ISO 15530-3, Geometrical product specifications (GPS) – Coordinate measuring machines (CMM): Technique for determining the uncertainty of measurement, Part 3: Use of calibrated workpieces or measurement standards, International Organization for Standardization (ISO), Central Secretariat, Geneva, Switzerland, 2011
9. ISO standards, International Organization for Standardization (ISO), Central Secretariat, Geneva, Switzerland. ISO standards are usually available from a national publisher, e.g. in Germany from Beuth Verlag, Berlin, a subsidiary of the German Institute for Standardization (DIN)
10. VDI/VDE 2617 Part 6.2: Accuracy of coordinate measuring machines. Characteristics and their testing. Guideline for the application of DIN EN ISO 10360 to coordinate measuring machines with optical distance sensors (2004)
11. Haitjema, H., Morel, M.A.A.: Noise bias removal in profile measurements. Measurement. **38**, 21–29 (2005)

References

Chapter 2
Tactile Surface Metrology

Abstract This Chapter is concerned with methods for characterization and measurement of technical surfaces using tactile methods. It comprises the classical stylus tip measurement as well as atomic force microscopy. Tactile surface profiling is approved since many decades as a reliable dimensional measuring technique. Atomic force microscopy is used in research, development, and quality assurance whenever high-resolution measurements of the surface structures are demanded.

2.1 Tactile Surface Profiling

The *tactile surface profiling* is approved since many decades and is a reliable measuring technique that is the best understood dimensional measurement. Stylus profilometry was developed a long time ago as a valuable means to study the surface roughness of materials [1, 2].

All tactile sensors operate on the principle of mechanical contact with the workpiece. Figure 2.1 shows a sketch of the working principle of a *stylus tip*.

A fine tip of mostly ruby or diamond (1) on a carrier (2) gets drawn horizontally over the surface (3). The stylus force is in the order of 1 mN. The size of the contact area amounts to 2–10 μm. While the tip follows the profile the carrier moves vertically (4). This vertical movement is transduced into an electrical signal either by a piezoelectric or an inductive transducer or an optical transducer. The electrical signal is then recorded as profile function $z(x)$. With a tactile sensor, the form and size of the probing form element as well as the spatial position and geometric form of the object surface to be measured are contained in the measured profile.

In principle all stylus sensors should be as rigid as possible to detect the interaction forces without deformation. The most often used tip shape is a sphere. Recommended radii are 2 ± 1 μm, 5 ± 2 μm, and 10 ± 3 μm. Recommended tip angles are 60°± 5° and 90°± 5°. Most frequently the tip has a radius of 5 μm and an aperture angle of 60°. For the stylus tip two properties are important: the accuracy in shape and the material. The deviation from the spherical shape is often specified in

M. Quinten, *A Practical Guide to Surface Metrology*, Springer Series in Measurement Science and Technology,
https://doi.org/10.1007/978-3-030-29454-0_2

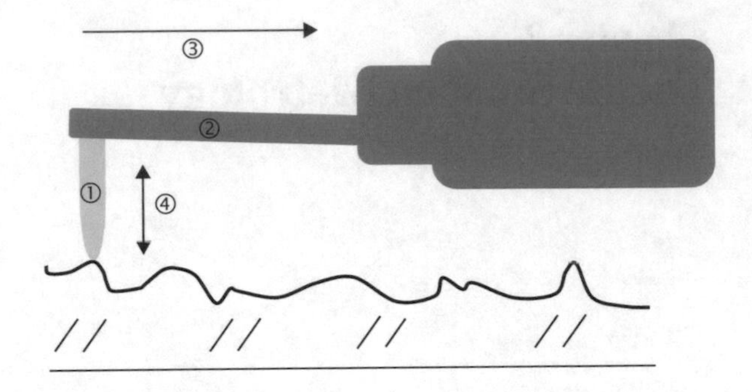

Fig. 2.1 Sketch of the working principle of a stylus tip

Table 2.1 Deviation of the tip shape from the spherical shape measured in grades

Grade	Deviation of the tip shape from spherical shape [μm]
3	0.08
5	0.13
10	0.25
16	0.4
20	0.5

Fig. 2.2 Typical stylus tip with spherical shape used in coordinate measuring machines (CMM). (Courtesy of Goekeler Messtechnik GmbH, Lenningen, Germany)

grades. Table 2.1 gives an overview on the grade and the belonging maximum deviation from the spherical shape.

Popular materials for the spherical tip are ruby, sapphire, alumina, silicon nitride, zirconia, steel, and diamond. The choice of the material depends on the application. A typical stylus tip with ruby sphere for coordinate measuring machines (CMM) is shown in Fig. 2.2.

The actual position of the probing point cannot be determined without a mathematical correction based on the known coordinates of the center point of the stylus tip. For an exact correction, the probing element must be calibrated or qualified carefully. Already in 1970 shortly after the invention of stylus profilometry the influence of the stylus tip radius on the roughness values was considered by Radhakrishnan [3]. As the size of the probing tip has influence on the measured profile it is normally necessary to probe a larger number of points on any geometric feature to be measured. Geometric structures smaller than the tip size cannot be

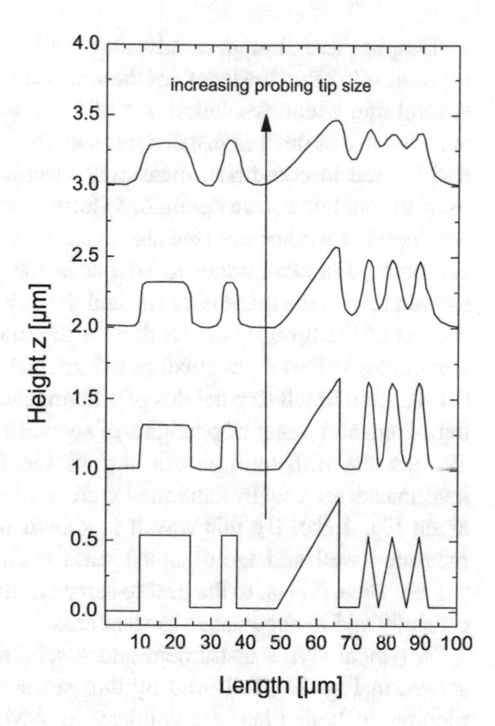

Fig. 2.3 Influence of the probing tip size on the measured profile. On bottom the original profile. The measured profiles are shifted along the ordinate by a constant value for better presentation

recorded. Steep edges are also a problem as they are not correctly recognized due to the high taper angle. The influence of the probing tip on the recorded profile is illustrated in Fig. 2.3.

The measuring rate of a stylus instrument is low (max. 50–500 µm/s, depending upon the surface) caused by the mechanical contact with the workpiece. The stylus force *F* of about 0.75–1 mN can already lead to ductile deformations and scratches on the surface of the workpiece as well as material from the workpiece is deposited on the tip or the tip abrades. According to Hertz [4] the static elastic deformation *d* of a material is

$$d = \frac{9 \cdot F^2}{16 \cdot R \cdot \hat{E}^2} \tag{2.1}$$

when using a spherical tip with radius *R*. Here, $\hat{E}$ is the reduced elastic modulus with

$$\frac{1}{\hat{E}} = \frac{1 - \kappa_{tip}^2}{E_{tip}} + \frac{1 - \kappa^2}{E} \tag{2.2}$$

with κ and *E* the Poisson ratio and Young's modulus of the material and κ_{tip} and E_{tip} the Poisson ratio and Young's modulus of the tip.

Despite these obvious disadvantages this technique also has many advantages. As it is possible to setup the length of the scanned profile the measuring range amounts up to several mm with a resolution in z of a few nm. Therefore, tactile measuring systems are available as devices that can measure the form and the surface (stylus instruments), and is used in coordinate measuring machines (CMM). The stylus instruments are used to monitor coarse deviations (form deviations) and fine deviations (roughness, waviness) of workpieces (see also again Sect. 1.2). The measurement with a stylus tip can be used in laboratories as well as in the production line. It is independent of the illumination at the measuring site and the color of the workpiece. However, due to the contact of the tip with the workpiece the material surface must be sufficiently hard. Roughness values R_q as small as 0.5 nm can be measured with a lateral resolution of 0.1–0.2 μm. Small step heights of 100 nm can be measured with a repeatability of less than 1 nm and larger step heights of several microns with a repeatability of 5–10 nm. To measure film thickness, a step of the film to the substrate is needed. Stylus instruments are usually calibrated with certified roughness standards (see for example again Fig. 1.6b). By this way it is shown that the surface of the standards can be measured well and the obtained surface characteristics coincide with the certified values. Then, it is up to the user to carry out the measurements on the relevant samples carefully and in accordance to standards.

A typical stylus instrument and a sensing arm for roughness measurement are shown in Fig. 2.4, followed by two surface roughness mappings in Fig. 2.5. The pictures in both Figs. are courtesy of AMETEK GmbH – Business Unit Taylor Hobson, Weiterstadt, Germany.

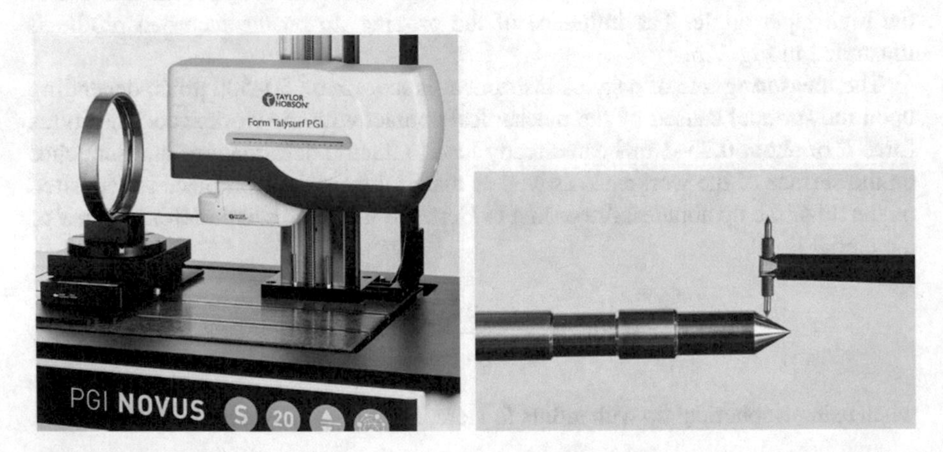

Fig. 2.4 Typical high precision stylus measurement system (left) and sensing arm for roughness measurement (right). (Courtesy of AMETEK GmbH – Business Unit Taylor Hobson, Weiterstadt, Germany)

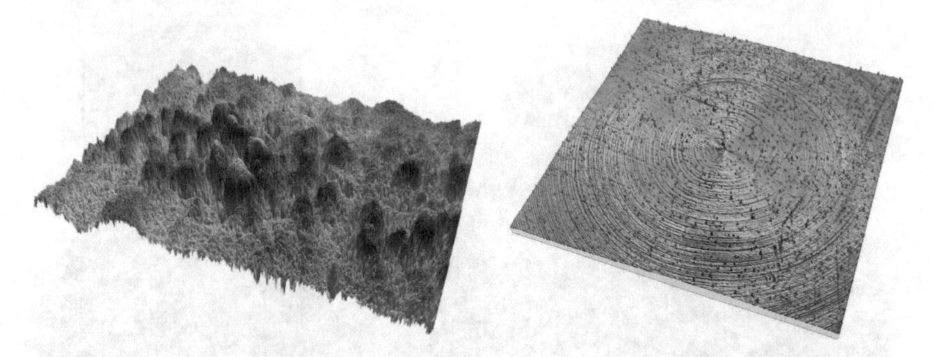

Fig. 2.5 Surface roughness mappings of two different surfaces with a stylus tip. (Courtesy of AMETEK GmbH – Business Unit Taylor Hobson, Weiterstadt, Germany)

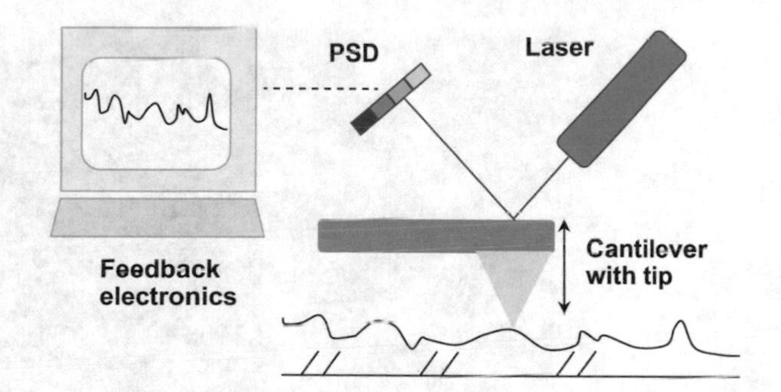

Fig. 2.6 Sketch of the principle of an atomic force microscope (AFM)

2.2 Atomic Force Microscopy

The *atomic force microscope* (AFM) was developed and first used by Binnig, Quate, and Gerber in 1985 [5]. One year later the Nobel Prize was granted to Binnig and Rohrer for the discoveries of scanning tunneling microscopy (STM) and AFM. From this timepoint AFM became commercially available. The method is increasingly used in research, development, and quality assurance whenever high-resolution measurements of the surface structures are demanded.

Atomic force microscopes scan the surface of a sample using a very fine tip at the end of a micro-scaled cantilever (see Fig. 2.6 for the principle of an atomic force microscope). Atomic interactions between the surface and the tip lead to a deflection of the cantilever. The deflection of the cantilever due to the atomic forces results in a displacement of the laser beam on a position-sensitive photodetector. For small enough bendings of the cantilever, the displacement of the beam is proportional to the force on the tip. A force of 10^{-10} N causes a deflection of 0.1 nm. Based on this beam deflection method, the AFM measures the 3D topography and nano roughness almost non-destructively with near-atomic resolution in contact and non-contact mode. The position-sensitive diode (PSD) can detect tip displacements smaller than 1 nm (up to 0.1 nm) as shown by Hues et al. [6].

Fig. 2.7 AFM pictures (**a**) compact system CoreAFM, (**b**) lens system LensAFM. (Courtesy of Nanosurf AG, Liestal, Switzerland)

The lateral measuring range is determined by the piezo scanner and is in the order of 100 μm. In vertical direction heights of several microns can be measured. AFM works on conducting as well as isolating surfaces. For the latter additional electrostatic forces between surface and tip may occur that can lead to errors in the measurement. AFM can be operated in principle without vacuum conditions but then the measurement is more complicated. The biggest disadvantage of the AFM is the relatively long measurement time. The maximum scan velocities amount to 10–50 μm/s caused by the properties of the cantilever.

Typical AFM setups are shown in Fig. 2.7. The pictures are courtesy of Nanosurf AG, Liestal, Switzerland. The two photographs in Fig. 2.7a show the compact tabletop AFM CoreAFM. For inserting the sample the system can be opened. The picture in Fig. 2.7b shows the AFM measuring head LensAFM. It can be mounted on a microscope similar to a microscope objective. This lens AFM is well-suited for measurements in surface metrology.

The key element of an AFM is the *cantilever*. It is the microscopic force sensor formed by one or more beams of silicon or silicon nitride of 100–500 μm length and

Fig. 2.8 Pictures of typical AFM cantilevers with tips. (© 2017 NanoWorld AG, Neuchâtel, Switzerland)

about 0.5–5 μm thickness. A sharp tip is mounted on the end of the cantilever which is used to sense a force between the sample and the tip. For common topographic measurements the probe tip is plain silicon or silicon nitride and is brought into continuous or intermittent contact with the sample and gets scanned over the surface. For specific applications tips can be coated with conductive, magnetic, or bio-active material.

To achieve high lateral resolution the tips have to be as sharp as possible which is best achieved with conical or pyramidal tips with the radius of their apex being in the order of 1–10 nm resulting in lateral resolutions of 1–10 nm. The tip height is pretty small with about 10–15 μm for which reason only small step heights can be measured. Also the apex angle of the tip of up to 70° must be taken into account particularly when measuring at steep edges.

Examples of cantilevers with tips are shown in Fig. 2.8. The pictures are courtesy of Nanoworld AG, Neuchâtel, Switzerland.

The second important concept in AFM operation is feedback. Any feedback parameter is tried to be kept constant at its setpoint using a proportional-integral-derivative control (PID). For AFM operation the integral gain is most important and can have a most dramatic effect on the image quality. The proportional gain might provide slight improvement after optimization of the integral gain. The derivative gain is mainly for samples with tall edges. Which parameter is used and controlled in the feedback loop depends upon the used operation mode (see below). Other parameters that are important in feedback are the scan rate and the actual value of the setpoint. If the scan rate is too fast the PID loop will not have sufficient time to adjust the feedback parameter to its setpoint value. Then, the measured height will deviate from the true topography particularly at slopes and near edges. A setpoint close to the parameter value out of contact feedback is most gentle for the sample but tends to slow down the feedback.

One distinguishes three operation modes, a *contact mode*, a *non-contact mode*, and a *tapping mode*. In the *contact mode* the tip is very close to the surface (about 1 nm) and the repulsive steric force between tip and surface leads to the bending of the cantilever. Contact mode is most useful for hard surfaces and is rather insensitive for specific sample characteristics, otherwise the tracking force may lead either to contamination from removable material on the surface or even to damages of the surface or to erosion of the sharpness of the probe tip. This mode is the most similar to a measurement with a stylus tip although the tip geometry is different and the tracking force is substantially smaller. In contact mode AFM the probe tip is scanned over the surface in a raster pattern. The feedback loop maintains a constant cantilever deflection and consequently a substantial, constant force on the sample.

In the *non-contact mode* long-range forces are used. They are detected via the change of the cantilever resonance frequency due to the gradient of the force. Typically, the cantilever is driven near the resonance frequency of the cantilever which is in the order of 20–300 kHz. The distance between the tip and the surface is considerably larger than in the contact mode. The resonance frequency of a spring or a cantilever is given by its spring constant *D* and its mass *m*

$$f = \frac{1}{2\pi}\sqrt{\frac{D}{m}} \tag{2.3}$$

The spring constant *D* can be calculated from the cantilever width *w*, the thickness *t*, the length *L* and Young's modulus *E* of the cantilever material as

$$D = \frac{E \cdot w \cdot t^3}{4 \cdot L^3} \tag{2.4}$$

Typical values of the spring constant for commercially available cantilevers range from 0.01 to 50 N/m. Equation (2.3) shows that a cantilever can have both low spring constant and high resonant frequency if it has a small mass. Therefore, AFM cantilevers tend to be very small. When approaching the surface, the resonance frequency changes as the long-range forces alter the spring constant by the additive amount D_{ts} resulting from the average tip-sample force gradient.

The origin of the long-range forces are quantum mechanical oscillations of the electrons of an atom around their position of rest. These oscillations cause a small electrical dipole at this atom. The dipole interacts with the electrons respectively a similar dipole at the neighboring atom. This dipole-dipole interaction is a very short range interaction with the force decreasing with the distance r between the atoms proportional to $1/r^6$. This force is called *van der Waals force*. In an atomic force microscope more than one atom of the tip and the surface are involved so that all van der Waals forces of all involved atoms must be considered to get the actual dependence on the distance. In the standard Hamaker approach for simple geometries this means to sum up all van der Waals forces. For example, for a spherical tip approaching a flat surface the resulting force is

$$F_{vdW} = -\frac{H \cdot R}{6 \cdot r^2} \tag{2.5}$$

with H being the Hamaker constant. Van der Waals forces are substantially weaker than the forces used in contact mode. Hence, this mode is more sensitive to environmental vibrations and capillary forces from thin liquid films on the sample in a normal atmosphere.

The most popular form of AFM operation mode is the *tapping mode* [7–9]. In this patented technique the tip is brought in contact with the surface, just as in the contact mode, but is then lifted again off the surface to avoid dragging the tip across the surface. This is enabled by mounting the cantilever into a holder with a shaker piezo. The shaker piezo allows for oscillation of the probe tip with frequencies between 100 Hz and 2 MHz. Typically, it is driven near the resonance frequency of the cantilever (20–300 kHz). This oscillation frequency is here also the feedback parameter. The piezo motion causes the cantilever to oscillate with a high amplitude of typically greater than 20 nm when the tip is not in contact with the surface. For measurement the oscillating tip is moved toward the surface until it begins to lightly touch or "tap" the surface. During scanning the vertically oscillating tip alternately contacts the surface and lifts off. As the oscillating cantilever begins to intermittently contact the surface, the cantilever oscillation is reduced due to energy loss caused by the tip contacting the surface. The reduction in oscillation amplitude is used to identify and measure surface features. Tapping mode overcomes major problems associated with friction, adhesion, electrostatic forces, and other tip-sample related difficulties. On retraction the pulling force must be sufficiently large to overcome adhesion to separate the surface and the tip. The adhesion has several causes. The most important among them are:

- Increase of the contact area due to the elastic deformation of the surface and the tip in the contact.
- Capillary forces due to the condensation of water from the ambient air in and around the contact region.
- Short-range chemical binding forces which contribute to the total adhesion energy.

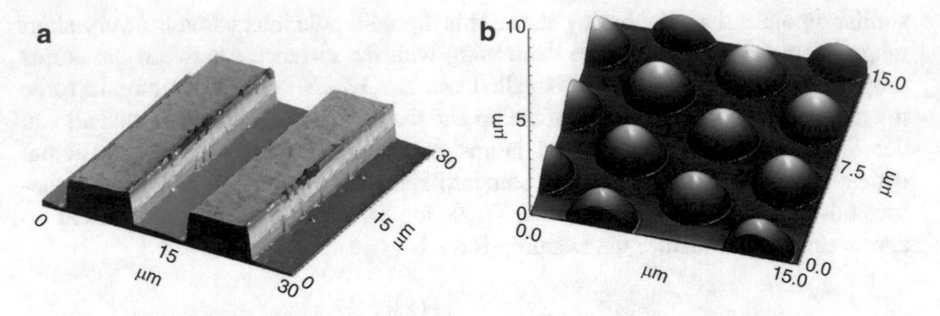

Fig. 2.9 AFM topography measurements: (**a**) two fingers of a photonic band gap filter, (**b**) surface topography of a sapphire wafer with PSS stuctures. Courtesy of FRT GmbH, Bergisch Gladbach, Germany

In tapping mode, the stiffness of the cantilever must be appropriately suited to image the material. Otherwise, damages of the sample or tip wear may occur if the cantilever is too stiff or the cantilever does not interact sufficiently with the sample to obtain high-contrast images if the stiffness is too low.

In Fig. 2.9 two exemplaric AFM topography measurements are shown. The pictures are courtesy of FRT GmbH, Bergisch Gladbach, Germany. The first picture in Fig. 2.9a shows a part of a photonic band gap filter (area 30 × 30 μm) and in Fig. 2.9b the topography of a section (area 15 × 15 μm) of a sapphire wafer with periodic structures on the sapphire surface is depicted. Sapphire wafers with periodic structures of various shapes such as cone, dome, pyramid, and pillar, etc., are called patterned sapphire substrates (PSS) and are used for enhancing LED light extraction.

The AFM allows probing of a variety of forces beyond van-der Waals forces, e.g. ion-ion repulsion forces, electrostatic forces, magnetic forces, capillary forces, adhesion, or frictional forces. Hence, several extensions of the AFM exist utilizing different kinds of interaction. They are briefly summarized below.

Force Modulation Microscopy (FMM)
In *Force Modulation Microscopy* the AFM is operated in the contact mode. The force on the sample is modulated such that the average force on the sample is equal to that in contact mode [10]. For this the cantilever oscillates in vertical direction with the oscillation being significantly faster than the raster scan rate. This is achieved with an additional piezo electric actuator on the cantilever holder. Under the same tracking force, a stiff area on the sample deforms less than a soft area. Stiffer areas put up greater resistance to the vertical oscillation of the cantilever and

consequently cause greater bending of the cantilever. The variation in cantilever deflection amplitude at the frequency of modulation is a measure of the relative stiffness of the surface.

Lateral Force Microscopy
Frictional (FFM) or *Lateral Force Microscopy* (LFM) is a technique that identifies and maps relative differences in surface frictional characteristics. With the lateral force technique the AFM is operated in contact mode but the probe is additionally scanned sideways on the fast axis and forward-back on the slow axis [11]. The torsion of the cantilever supporting the probe will increase or decrease depending on the stiffness and adhesion arising from various regions of the surface (greater torsion results from increased friction). The four-quadrant diode (the PSD) can simultaneously measure and record topographic data and from the lateral force data elastic and viscoelastic properties.

Chemical Force Microscopy
Chemical Force Microscopy (CFM) is a special kind of Lateral Force Microscopy where the tip is functionalized with a chemical species and is scanned over a sample to detect adhesion differences between the species on the tip and those on the surface of the sample. Molecules bound to the AFM probe can be used as chemical sensors to detect forces between molecules on the tip and target molecules on the surface.

Scanning Surface Potential Microscopy, Kelvin Probe Microscopy
With this extension the work function of surfaces can be observed at atomic scales. The work function is related to many surface phenomena, the most prominent being catalysis, surface reconstruction, doping and band bending of semiconductors, and corrosion. Mapping the work function with Kelvin probe microscopy gives information about the composition and electronic state of the local structures on the surface of a solid.

The method is based on the measurement of the electrostatic forces between the small AFM tip and the sample [12]. The conducting tip and the sample are characterized by (in general) different work functions which represent the difference between the Fermi level and the vacuum level for each material. Bringing the tip and the surface into contact a net electric current flows between them until the Fermi levels are aligned. When the AFM tip is brought only close to the sample surface an electrical force is generated between the tip and sample surface, due to the differences in their Fermi energy levels:

$$e_0 \cdot U = \Phi_{\text{sample}} - \Phi_{\text{tip}} \tag{2.6}$$

where Φ_{tip} and Φ_{sample} are the work functions of the tip and the sample.

Magnetic Force Microscopy MFM

For this application AFM-tips are coated with magnetic materials [13–15]. Two pictures of the sample are taken. First a topographic picture is acquired in the tapping mode. Then, the tip is raised to a lift scan height where the second picture in the tapping mode is acquired. This second picture is obtained due to the spatial variation of the long-range magnetic forces.

Electrostatic Force Microscopy EFM

Using conductive AFM-tips an electrical field can be applied that is very sensitive on electrical fields of the sample [16, 17]. Similar to the MFM here also two pictures are acquired from the sample in the tapping mode, one topographic picture and one at the lift scan height caused by the long-range electrostatic Coulomb forces.

References

1. Williamson, J.P.B.: The microtopography of surfaces. Proc. Inst. Mech. Eng. London. **183**(3K), 21–31 (1967/1968)
2. Teague, E.C., Scire, F.E., Baker, S.M., Jensen, S.W.: Three-dimensional stylus profilometry. Wear. **83**, 1–12 (1982)
3. Radhakrishnan, V.: Effect of stylus radius on the roughness values measured with tracing stylus instruments. Wear. **16**, 325–335 (1970)
4. Hertz, H.: H., Ueber die Beruehrung fester elastischer Koerper, J. Angew. Math. **92**, 156–171 (1881)
5. Binnig, G., Quate, C.F., Gerber, C.: Atomic force microscope. Phys. Rev. Lett. **56**(9), 930–933 (1986)
6. Hues, S.M., Colton, R.J., Meyer, E., Güntherodt, H.-J.: Scanning probe microscopy of thin films. MRS Bull. **18**, 41–49 (1993)
7. Zhong, Q., Inniss, D., Kjoller, K., Elings, V.B.: Tapping mode atomic force microscopy. Surf. Sci. Lett. **290**, L688–L692 (1993)
8. Cleveland, J.P., Anczykowski, B., Schmid, A.E., Elings, V.B.: Energy dissipation in tapping mode atomic force microscopy. App. Phys. Lett. **72**, 2613–2615 (1998)
9. B. Virgil, J. A. Gurley, Tapping atomic force microscope with phase or frequency detection, U.S. Patent RE36,488 (2000)
10. Maivald, P., Butt, H.J., Gould, S.A.C., Prater, C.B., Drake, B., Gurley, J.A., Elings, V.B., Hansma, P.K.: Using force modulation to image surface elasticities with the atomic force microscope. Nanotechnology. **2**, 103–106 (1991)
11. Meyer, G., Amer, N.M.: Simultaneous measurement of lateral and normal forces with an optical-beam-deflection atomic force microscope. Appl. Phys. Lett. **57**(20), 2089–2091 (1990)
12. Nonnenmacher, M., Oboyle, M.P., Wickramasinghe, H.K.: Kelvin probe force microscopy. Appl. Phys. Lett. **58**(25), 2921–2923 (1991)
13. Sáenz, J.J., García, N., Slonczewski, J.C.: Theory of magnetic imaging by force microscopy. Appl. Phys. Lett. **53**, 1449–1454 (1988)
14. Martin, Y., Wickramasinghe, H.K.: Magnetic imaging by “force microscopy” with 1000 Å resolution. Appl. Phys. Lett. **50**, 1455–1457 (1987)

15. Rugar, D., Mamin, H.J., Guethner, P., Lambert, S.E., Stern, J.E., McFayden, I., Yogi, T.: Magnetic force microscopy: general principles and application to longitudinal recording media. J. Appl. Phys. **68**, 1169–1183 (1990)
16. Martin, Y., Abraham, D.W., Wickramasinghe, H.K.: High-resolution capacitance measurement and potentiometry by force microscopy. Appl. Phys. Lett. **52**, 1103–1105 (1988)
17. Stern, J.E., Terris, B.D., Mamin, H.J., Rugar, D.: Deposition and imaging of localized charge on insulator surfaces using a force microscope. Appl. Phys. Lett. **53**, 2717–2719 (1988)

Chapter 3
Capacitive and Inductive Surface Metrology

Abstract Interactions among a surface and a probing unit are not restricted on mechanical forces. This became evident already from the extensions of the atomic force microscope in the previous chapter. Strong interactions can also occur for long-range electrical and magnetic forces. This Chapter deals with the most common surface metrology methods based on electrical fields -the capacitive surface profiling- as well as magnetic fields -the profiling with eddy currents.

3.1 Capacitive Surface Profiling

Electrical fields play a role in surface metrology mainly in capacitive methods. Capacitive methods base upon the variation of the capacitance of an capacitor with the distance.

For a capacitor with two parallel plates as illustrated in Fig. 3.1a the capacitance C is given as

$$C = \varepsilon_0 \cdot \frac{A}{d} \tag{3.1}$$

with A being the area of the plates and d the distance between the plates. $\varepsilon_0 = 8.854 \cdot 10^{12}$ As/Vm is a natural constant.

If a dielectric material is put in between the plates as illustrated in Fig. 3.1b the capacitance increases by a factor ε

$$C = \varepsilon \cdot \varepsilon_0 \cdot \frac{A}{d}. \tag{3.2}$$

The reason is that the electric field E between the two plates induces a dipole p at each atom. They sum up to a net polarizability P that adds to the electric field E.

A capacitance measurement system in surface metrology consists of a parallel plate capacitor with the sensor as one plate and the target being measured as the other plate. The sensor plate is a shielded electrode in a probe head as shown Fig. 3.2.

M. Quinten, *A Practical Guide to Surface Metrology*, Springer Series in Measurement Science and Technology,
https://doi.org/10.1007/978-3-030-29454-0_3

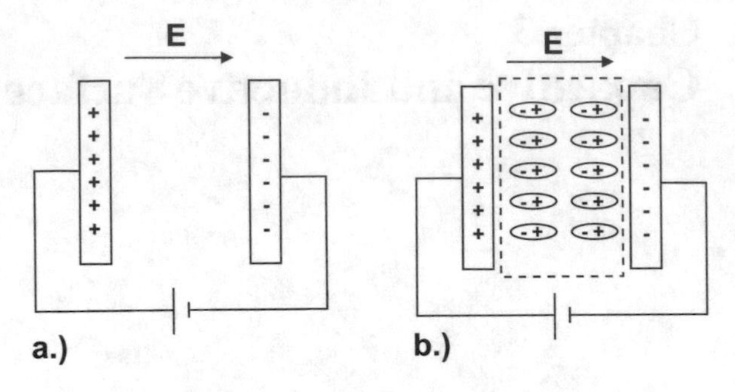

Fig. 3.1 (**a**) Empty capacitor, (**b**) Capacitor filled with a dielectric material

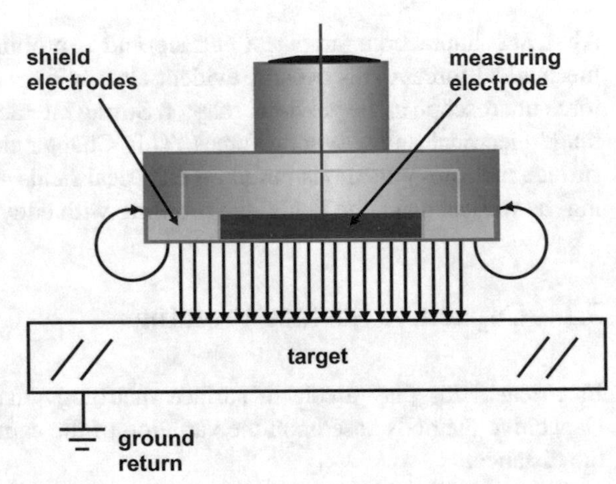

Fig. 3.2 Sketch of a shielded measuring electrode in a probe head as used in surface metrology

Then, the electric field between the sensor plate and an ideally plane target is very homogeneous. The electric flux lines of the shield electrodes are not considered in the measurement.

To create the electric field between the two plates the target must be conductive. The composition or thickness of the target is unimportant. To complete the capacitance circuit, the target should be grounded back to the amplifier. For optimal performance a conductive path is required. Capacitive coupled targets can work well if the capacitance is 0.01 μF or higher.

The lines of flux in the electric field established between the probe and target always leave the measuring electrode normal to its surface and always enter the target normal to its surface (see Fig. 3.2). If the sample is clearly larger than the sensor the electric field within the sensing area is consistent and linear. Field distortions may occur at the borders of the target or if the sample is too small. The field distortion will create measurement errors by degrading the sensor linearity and changing its measurement range.

As capacitance sensors have a relatively large sensing area in relationship to their measurement range they take an average measurement to the surface in question. Then, features of the surface that are smaller than the sensing element may not be detected or the sensor output may not respond accordingly. Therefore, roughness of

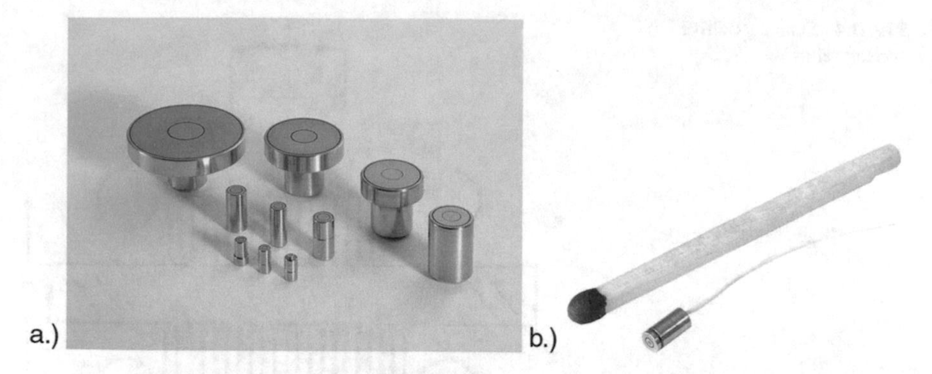

Fig. 3.3 (**a**) Standard capacitive sensor types, (**b**) World's smallest capacitive sensor. (Pictures are courtesy of E + H Metrology GmbH, Karlsruhe, Germany, www.eh-metrology.com)

the specimen is automatically averaged. A few attempts were made to assess also the surface roughness of machined parts using a capacitive sensor [1–5]. The probe size affects the sensor output similar to the probing tip size in tactile profiling (see again Fig. 2.3). Because of this, the probe tip should always be 25% smaller than the smallest feature targeted for measurement.

Similar to optical sensors the capacitive sensors work contactless. The measurement range varies typically between 0.05 mm and 10 mm. Capacitance sensors offer extremely high resolution in z-direction (0.0005% of the measuring range, up to 25 nm) often exceeding that of complex laser interferometer systems. The measurement signal is pretty stable with a linearity between 0.025% and 0.05% of the measurement range. Digital correction typically yields a linearity of ±0.01% or better. The primary factor in determining resolution is the electrical noise in the system. The robust electrodes can be used in extreme environments as well as in presence of magnetic fields. They are insensitive to ultrasonic noise, electromagnetic fields, lighting conditions, and humidity. The most common environmental problem that can affect the accuracy of a capacitive sensor is temperature. In Fig. 3.3 a set of pictures of capacitive sensors is shown. The pictures are courtesy of E + H Metrology GmbH, Karlsruhe, Germany, www.eh-metrology.com.

Capacitive sensors can also be used to measure the thickness of conducting materials such as semiconductors and nonconducting materials such as plastics, foils, quartz, glass, ceramics, etc. For this, preferably a setup with two opposite capacitive sensors with same detector area is used as illustrated in Fig. 3.4.

For conductive materials like silicon the system acts as two capacitors in series because the conductive target material does not contribute to the total capacitance. If the distance between the two probes is D, the thickness of the sample is t, and the distances of the two probes to the target are d_1 and d_2, the total capacitance is obtained as

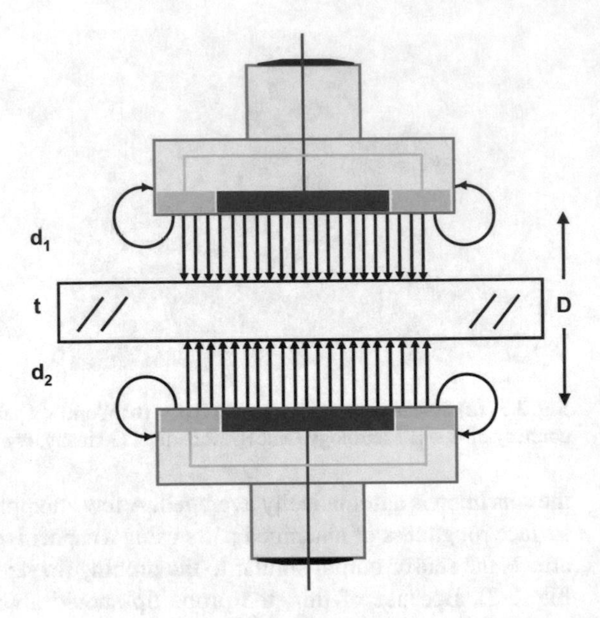

Fig. 3.4 Dual capacitive sensor setup

$$\frac{1}{C_{total}} = \frac{D - t}{\varepsilon_0 A} \tag{3.3}$$

since $d_2 = D\text{-}t\text{-}d_1$. This relation can be resolved for the thickness t of the sample:

$$t = D - \frac{\varepsilon_0 A}{C_{total}}. \tag{3.4}$$

With this setup it is possible to measure not only the thickness of semiconductors but also other quantities that are relevant in the semiconductor industry, e.g. bow, warp, and total thickness variation (TTV). As an example Fig. 3.5 shows a capacitive surface mapping of a silicon wafer. The pictures represent the measurements (a) of the topography from top, (b) of the topography from bottom, and (c) of the warp of the wafer. The pictures are courtesy of E + H Metrology GmbH, Karlsruhe, Germany, www.eh-metrology.com.

For nonconducting materials it must be considered that the sample contributes with its capacitance. The additional capacitance of the sample amounts to

$$C_s = \varepsilon_s \cdot \varepsilon_0 \cdot \frac{A}{t} \tag{3.5}$$

with ε_s being the dielectric constant or relative permittivity of the sample material. Then, the total capacitance of three capacitors in series is

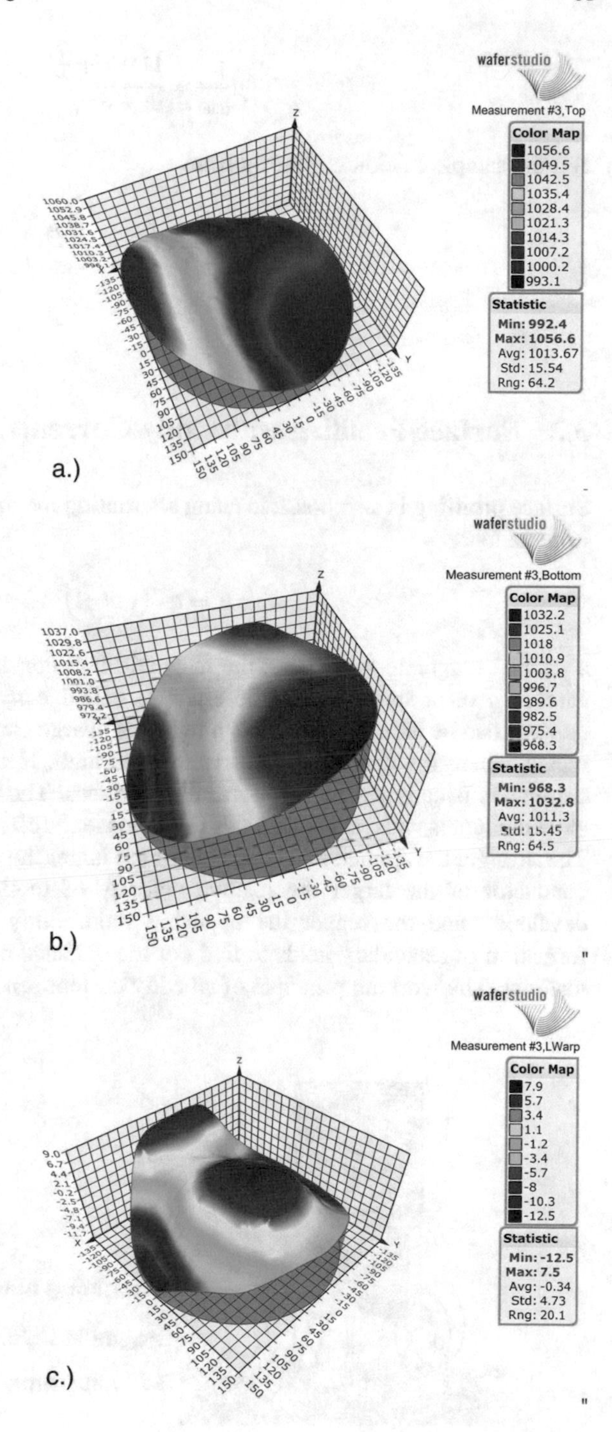

Fig. 3.5 Capacitive surface mapping of a silicon wafer: **a** topography from top, **b** topography from bottom, **c** the warp of the wafer. (Courtesy of E + H Metrology GmbH, Karlsruhe, Germany, www.eh-metrology.com)

$$\frac{1}{C_{total}} = \frac{D - t + \frac{t}{\varepsilon_s}}{\varepsilon_0 A} \tag{3.6}$$

and the sample thickness follows from

$$t = \frac{\varepsilon_s}{\varepsilon_s - 1} \cdot \left(D - \frac{\varepsilon_0 A}{C_{total}} \right). \tag{3.7}$$

3.2 Surface Profiling with Eddy Currents

Surface profiling is also possible using alternating magnetic fields. According to the Lorentz force

$$\mathbf{F} = q \cdot (\mathbf{v} \times \mathbf{B}) \tag{3.8}$$

a parallel magnetic field (here: the magnetic flux density $\boldsymbol{B}$) forces moving charge carriers q on a spiral path. This effect is called *eddy currents*. Noticeable eddy currents can be induced only if the number of charge carriers is high as in the case of metals where the density of free electrons is high. Hence, the application of eddy currents is usually restricted on metallic surfaces. The eddy currents in turn create electromagnets with magnetic fields that oppose the effect of applied magnetic field. The stronger the applied magnetic field or the larger the electrical conductivity of the conductor or the larger the relative velocity of motion the greater the currents developed and the bigger the opposing field. Eddy current probes senses this formation of secondary fields to find out the distance between the probe and target material. The working principle of an eddy current sensor is depicted in Fig. 3.6.

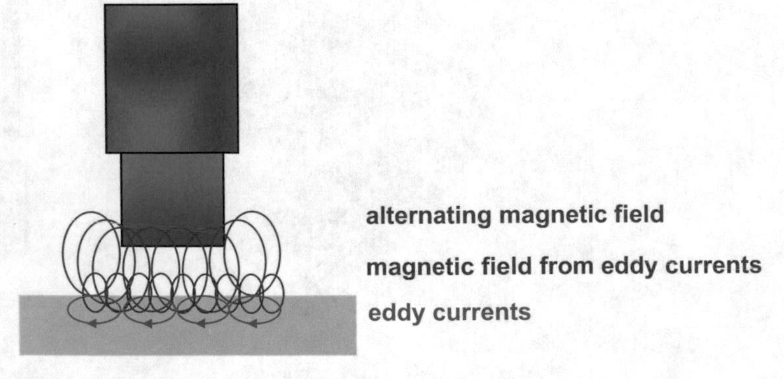

Fig. 3.6 Working principle of an eddy current sensor

The probe generates an alternating magnetic field (blue lines) which penetrates into the workpiece and induces eddy currents (green lines). The eddy currents in turn create an alternating magnetic field (red lines) that opposes the applied field.

The physical mechanisms in an eddy current sensor are best described by the complex impedance Z of the sensor coil represented as a series LR circuit

$$Z(\omega) = R + i\omega \cdot L \tag{3.9}$$

where ω is the operating frequency of the sensor in radians per second and $i = \sqrt{-1}$ is the imaginary unit. More directly connected to the ultimate performance of the sensor is its quality factor Q, defined as

$$Q(z) = \frac{\omega \cdot L(z)}{R(z)} \tag{3.10}$$

Q depends on the initial gap size, the standoff z_0. A high Q leads to high accuracy and stability. Practical operating frequency values for air core coils lie between 100 kHz and 10 MHz. Both, inductance L and resistance R change with the target position respectively the gap size. L decreases and R increases when reducing the gap size, and vice versa for increasing the gap. This is the case for nonmagnetic metals. For magnetic metals the inductance increases when reducing the gap size. The reason is that the magnetic permeability of the magnetic metal concentrates the magnetic field of the coil. This effect is stronger than the effect through the magnetic field of the eddy currents.

Having placed the probe on an initial gap size, changes in the metal thickness or defects like cracks will interrupt or alter the amplitude and pattern of the eddy currents and the resulting magnetic field. This in turn affects the movement of electrons in the coil, decreases the resistance of the coil and increases the inductance from their initial values, both for magnetic and nonmagnetic metals.

As the changes in magnitude and phase of the complex impedance Z often are very small, a distinct graphical representation was developed, the *normalized impedance plane*. In this plane the imaginary part of Z is plotted versus the real part of Z, both normalized to $|Z_0|$, the modulus of the coil impedance in air. The normalized impedance plane is shown in Fig. 3.7.

The blue line indicates the undisturbed dependence of the coil impedance upon the conductivity σ. For a metal with distinct conductivity the operating point (black dot) is set with the frequency ω. Any defect or changes in the metal thickness will alter the amplitude and phase of Z in a distinct way, leading to deviations from the blue line. They are indicated in red color. The meaning of the deviations is: A – sensor lifting, B - increase of relative permeability, C – inner defects, D – cracks, and E – inhomogeneities of the conductivity.

Since eddy currents tend to concentrate at the surface of a material, they can be used to detect surface and near-surface variations. The reason is the short penetration depth of magnetic (as well as electrical) fields in a metal, the so-called skin depth. The penetration depth obeys the relation

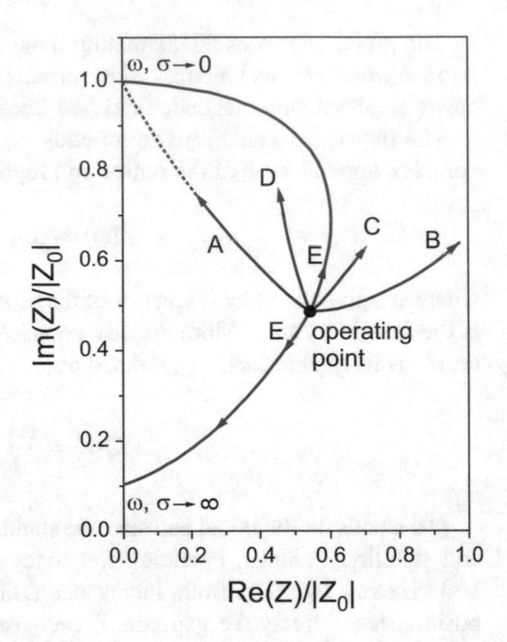

Fig. 3.7 The normalized impedance plane of an eddy current sensor. The black point defines the working point. The blue line indicates the undisturbed dependence of the coil impedance upon the conductivity. Deviations from the blue line are indicated in red. For their description see text

$$\delta = \frac{1}{\sqrt{0.5 \cdot \omega \cdot \mu \cdot \sigma}}. \tag{3.11}$$

The parameter μ is the magnetic permeability of the metal.

The higher the conductivity σ is the more sensitive is the eddy current sensor but the less is the penetration depth. This effect is also frequency dependent. For high frequencies ω a high near-surface resolution is obtained but with less penetration depth. These findings make eddy current sensors to a useful tool for near-surface crack detection, detection of corrosion, and for measuring the thickness of paints or other coatings on metallic substrates. In addition, determination of positions and gaps as well as conductivity measurements are fields of application. The variety of tests using eddy current instruments depends upon the type of probe. Therefore, a careful probe selection will help to optimize the performance.

Eddy current sensors for industrial applications have measuring ranges between 0.4 and 80 mm with linearities of 0.2% of the measuring range and a resolution of 0.005% of the measuring range. Practical eddy current sensors vary in diameter from a few millimeters to a meter and have maximum sensing ranges roughly equal to the radius of the coil. Linearity is typically 1% of the sensing range and noise levels are pretty low. A small sensor can resolve nanometer-size displacements with up to 10 μm total accuracy. They can be used in contaminated environments because they are neither sensitive to oil, dirt, water, or electromagnetic interference, etc., nor they are sensitive to (isolating) material in the gap between probe and target. On the other hand, if extremely high lateral resolutions are requested they are not well-suited. Also if large gaps between probe and target are requested they are not a good choice.

An eddy current system must always be calibrated with appropriate reference standards. Proper calibration is an essential part of any eddy current test procedure. For a normal calibrated operation the target surface must be at least three times larger than the probe diameter. Otherwise, a special calibration may be necessary.

For thickness measurement of paints or other nonmagnetic coatings on substrates of magnetic materials (Fe, Ni) another setup than an eddy current sensor is more appropriate. This magneto-inductive sensor consists of two separate coils on an iron core. One coil is used as field coil to generate an alternating magnetic field. When approaching the sensor to the film the magnetic field gets enhanced by the magnetic substrate and induces a voltage in the second coil which is dependent upon the distance. When putting the sensor on the film surface the distance corresponds to the film thickness and the measured voltage can be used to determine the thickness quantitatively.

References

1. Sherwood, K.E., Crookall, J.K.: Surface finish assessment by electrical technique. Proc. Inst. Mech. Eng. E (London). **182**(3K), 344–349 (1967/1968)
2. Brecker, H.N., Fromson, R.N., Shum, L.Y.: A capacitance based surface texture measuring system. Ann. CIRP. **25**(1), 375–377 (1977)
3. Lieberman, A.G., Vorburger, T.V., Giauque, C.H.W., Risko, D.G., Resnick, R., Rose, J.: Capacitance versus stylus measurement of surface roughness. Surf. Topogr. **1**, 315–330 (1988)
4. Williams, R.E., Rajurkar, K.P., Bishu, R. R.: Experimental comparison of a stylus based and a capacitance based surface roughness measurement system for different micro surface contour, Society of Manufacturing Engineers IQ990–255, 1–13 (1990)
5. Chang, H.-K., Kim, J.-H., Kim, I.H., Jang, D.Y., Han, D.C.: In-process surface roughness prediction using displacement signals from spindle motion. Int. J. Mach. Tools Manuf. **47**(6), 1021–1026 (2007)

Chapter 4
Optical Surface Metrology – Physical Basics

Abstract Surface metrology using optical methods is the field with the largest diversity since the interaction of light or electromagnetic radiation with solid matter is manifold. Therefore, it seems appropriate to have first an introduction into the physical basics of optics and optical sensors. This Chapter concentrates on the physical properties of electromagnetic waves, the basic interactions of light with matter, and on optical material properties. For more details than given in the sub-sections below the reader is referred to several common monographs and text books in optics and electrodynamics Stratton JA, Electrodynamic Theory, McGraw Hill, New York, 1941; Jackson JD, Classical Electrodynamics, 3rd edn. Wiley, New York, 1998; Born M, Wolf E, Principles of Optics, 7th edn. Cambridge University Press, Cambridge, 1999; Bergmann L, Schäfer C, Lehrbuch der Experimentalphysik, Vol. 3, Optik, 10th edn. W. de Gruyter, Berlin, 2004; Pedrotti F, Pedrotti L, Bausch W, Schmidt H, Optik für Ingenieure, Grundlagen, 4th edn. Springer, Berlin, 2007; Hecht E, Optics, 5th edn. Pearson Education Limited, Harlow, 2016.

4.1 Electromagnetic Waves

Light or in general *electromagnetic radiation* can be described with the model of a *wave* that propagates with the velocity of light c in a straight line in the vacuum. A wave is in general a process that is periodic in space and time, i.e. $W=W(\mathbf{r}, t)$. That means there is a *time period T* after which the wave W looks the same as at the timepoint t, i.e. $W(\mathbf{r}, t+T)=W(\mathbf{r}, t)$. The reciprocal value $1/T$ defines the *frequency* ν of the wave. Similar holds for the three-dimensional space. There is a periodicity $\mathbf{R}$ after that the wave appears the same as at a certain point $\mathbf{r}$, i.e. $W(\mathbf{r}+\mathbf{R}, t)=W(\mathbf{r}, t)$. The modulus $|\mathbf{R}|$ of this periodicity vector is the *wavelength* λ, defining the distance between two wave peaks.

The entity of electromagnetic radiation is illustrated in Fig. 4.1. The spectral range accessible to the human eye is called *light* and is rather small. It is indicated in Fig. 4.1 by the colored region in the mid of the graphics.

M. Quinten, *A Practical Guide to Surface Metrology*, Springer Series in Measurement Science and Technology,
https://doi.org/10.1007/978-3-030-29454-0_4

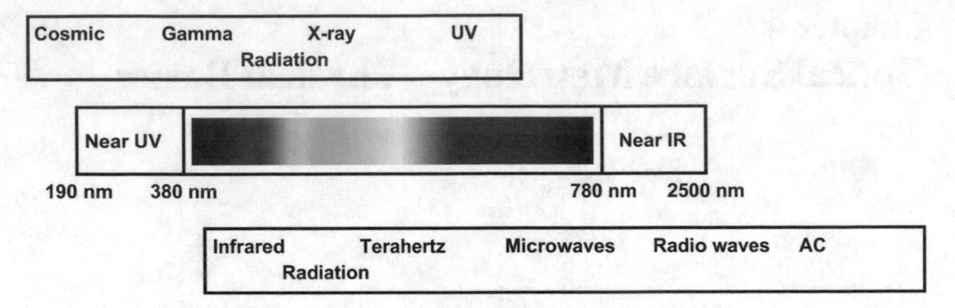

Fig. 4.1 The complete spectral range of electromagnetic radiation. The colored region from 380 to 780 nm is the visible spectral range

The spectral ranges are usually specified in different units depending on the application. Specifications in use are

electron volt (eV)	for the energy E of a radiation quantum, commonly used for visible light and radiation of higher frequency (UV, X-ray, . . .),
wavelength (nm)	for the wavelength λ of the radiation, commonly used for UV, visible light, and near infrared,
wavelength (μm)	for the wavelength λ of the radiation, commonly used for near and mid infrared, sometimes also for visible light,
wavenumber (cm^{-1})	for the reciprocal wavelength $\tilde{\nu} = 1/\lambda$, commonly used for near to far infrared, sometimes also for visible light.

Often, the *angular frequency* ω (1/s) is used instead of the frequency ν (Hz). They are related to each other via

$$\omega = 2\pi \cdot \nu. \tag{4.1}$$

All the quantities above can be converted from one to another. For this the following relations must be known:

$$\lambda \cdot \nu = c \tag{4.2}$$

for wavelength λ, frequency ν, and velocity of light c, and

$$E = h\nu = \hbar\omega = \frac{h \cdot c}{\lambda} \tag{4.3}$$

for the energy E of a radiation quantum of frequency ν respectively wavelength λ. Additionally the values of the natural constants Planck's constant $h = 6.62606896 \cdot 10^{-34}$ Js, $\hbar = h/(2\pi)$, velocity of light in vacuum

$c = 2.99792458 \cdot 10^8$ m/s, and elementary charge $e_0 = 1.60217648 \cdot 10^{-19}$ As (according to PTB and NIST), and the value of $\pi = 3.14159265359$ must be known. Then

$$E\ [\text{eV}] = \frac{1.23984187542}{\lambda\ [\mu\text{m}]} = \frac{1239.84187542}{\lambda\ [\text{nm}]}, \tag{4.4}$$

$$\omega\ \left[10^{15}\text{s}^{-1}\right] = \frac{1.88365156731}{\lambda\ [\mu\text{m}]} = \frac{1883.65156731}{\lambda\ [\text{nm}]}, \tag{4.5}$$

$$\tilde{\nu}\ \left[\text{cm}^{-1}\right] = \frac{10000}{\lambda\ [\mu\text{m}]} = \frac{10^7}{\lambda\ [\text{nm}]}. \tag{4.6}$$

When talking about electromagnetic waves, it is not mandatory, but often to find that a *harmonic wave* in time and space is given by

$$W(\mathbf{r}, t) = W_0 \cdot \exp\left(i \cdot (\mathbf{kr} - \omega t + \phi)\right) \tag{4.7}$$

with $\boldsymbol{k}$ being the *wavevector* that describes the propagation direction of the wave. Its modulus is $k = 2\pi/\lambda$. The quantity ϕ is an arbitrary constant. The complete expression $\boldsymbol{k} \cdot \boldsymbol{r} - \omega \cdot t + \phi$ is the *phase* of the wave. For an *electromagnetic* wave an *electric field* $\boldsymbol{E}(\boldsymbol{r},t)$ and a *magnetic field* $\boldsymbol{H}(\boldsymbol{r},t)$ must be considered that can be described by Eq. (4.7) and must fulfill Maxwell's equations. This description with complex numbers enormously simplifies the calculation of electromagnetic fields as well as further related quantities. For a brief introduction in the numerics with complex numbers see the Appendix. Fig. 4.2 depicts a harmonic electromagnetic wave.

As in vacuum by definition does exist nothing, no pertubation of the propagation of light by interaction with matter can occur. Yet, already one single hydrogen atom can destroy this harmony. The reasons are the force fields of the electromagnetic radiation: the electric field $\boldsymbol{E}$ and the magnetic field $\boldsymbol{H}$. On the other hand, an atom consists of electrically neutral (*neutron*) or positively charged (*proton*) tiny components that form the core of the atom, surrounded by negatively charged electrons that orbit the core like planets the sun (*Bohr's*

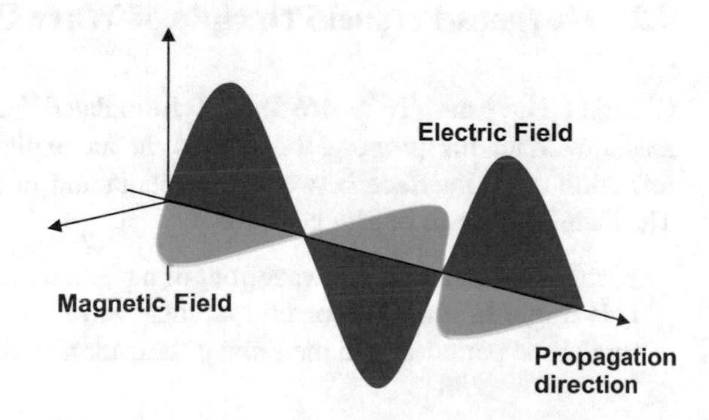

Fig. 4.2 Sketch of a harmonic electromagnetic wave

atom model) having a mass of about 1/2000 of the proton mass. There are as much electrons as protons to make the atom electrically neutral.

An electrical charge q in an electric field experiences a force $\boldsymbol{F_e}$ proportional to the electric field $\boldsymbol{E}$ and the size of the charge. In a magnetic field $\boldsymbol{H}$ the charge experiences a second force $\boldsymbol{F_m}$ proportional to the magnetic field and the size of the charge but perpendicular to the direction of movement if the charge is moving. The magnetic force is about 400 times smaller than the electric force because the magnetic field of an electromagnetic wave is about 400 times smaller than the electric field.

Caused by these forces the center of gravity of the positively charged protons separates from the center of gravity of the negatively charged electrons as both charges move in opposite direction. Actually, the protons are as heavy as under normal conditions of light ($|\boldsymbol{E}| < 100$ V/m) mainly the electrons get displaced. The charge separation makes the atom to an electrical *dipole* $\boldsymbol{p}=q\cdot\boldsymbol{r}$ (two opposite charges of same size with distance $\boldsymbol{r}$ between the centers of gravity). As the fields of an electromagnetic wave periodically change in time also the dipole changes periodically in time. That means that within a period of T the dipole changes from plus-minus to minus-plus. Hence, (mainly) the electrons experience a permanent acceleration by the electric and magnetic force. The energy contained in this acceleration (from work=force x path, $W=q\cdot\boldsymbol{E}\cdot\boldsymbol{r}=\boldsymbol{p}\cdot\boldsymbol{E}$) would destroy the atom within a short time for which reason the dipole partly reemits energy as an electromagnetic wave. Another part of this energy gets transformed into oscillations of the atoms around their position of rest. If these oscillations are periodic in time and space they are called *phonons*. They lead to a warming up of the specimen. In any case this part of energy has been retracted from the electromagnetic wave. This process is called *absorption*.

These facts are the microscopic basics for all phenomena related with electromagnetic waves and their interaction with matter. For a practical understanding of the various methods used in the surface metrology these basics are unsufficient and must be completed with further crucial principles, properties of electromagnetic waves, and more detailed descriptions.

4.2 Huygens-Fresnel Principle of Wave Propagation

Christian Huygens (1629–1695) first formulated how a wave propagates in a medium. With his proposal he arrived in an explanation of the reflection and refraction at an interface between two media and in an explanation of dichroism. The main statements of Huygens are

- Each point of an existing wavefront of a plane wave at the site $\boldsymbol{r}$ and at timepoint t_0 is origin of a new secondary circular wave with the same wavelength λ, the same time period T, and the same polarization as the original plane wave.

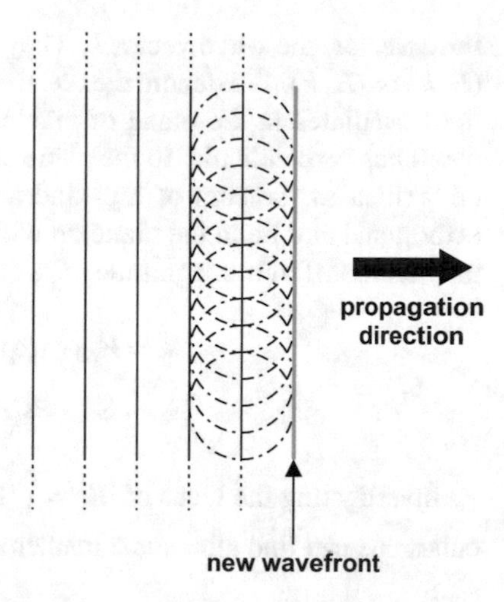

Fig. 4.3 Huygens' principle applied on a plane wave

- The envelope of these secondary waves determines the new wavefront and the stimulus in a point P.

The principle is illustrated in Fig. 4.3 for a plane wave. The black lines indicate the planes of constant phase where the magnitude of the wave is maximum. The dashed lines are the secondary circular waves. As it becomes obvious the envelope of the secondary waves form a new plane of constant phase at $\boldsymbol{r}+\lambda\boldsymbol{e}$, but also at $\boldsymbol{r}-\lambda\boldsymbol{e}$, with $\boldsymbol{e}$ being the unit vector in propagation direction. The latter is a gap in Huygens' explanation because it cannot bc excluded that the incident wavefront gets weakened by the secondary waves.

Auguste Jean Fresnel (1788–1827) closed this gap when combining the ideas of Huygens with Young's principle of interference by linear superposition (Thomas Young, 1773–1829). From this it follows that the envelope of the secondary waves results in a new wavefront only in propagation direction. One further important result of this combination is the statement

- The propagation direction of the reconstructed wave is perpendicular to the wavefront.

With the *Huygens-Fresnel principle* not only the straight propagation of waves can be described but also all deviations from it which are known as reflection, refraction, and diffraction.

4.3 Polarization

Electromagnetic waves (but also other waves) exhibit a property which is characteristic for transverse waves: *polarization*. For transverse waves the electromagnetic fields $\boldsymbol{E}$ and $\boldsymbol{H}$ are perpendicular to the propagation direction of the wave, given by a third vector, the wave vector $\boldsymbol{k}$. They oscillate in a fixed plane spanned either by $(\boldsymbol{E}, \boldsymbol{k})$ or $(\boldsymbol{H}, \boldsymbol{k})$. We denote the electromagnetic wave as *p-polarized* if the electric field oscillates in the plane of incidence, and as *s-polarized* if the electric field oscillates perpendicular to the plane of incidence. In general, the electric field can be written as the sum of a p- and a s-polarized component $\boldsymbol{E_p}$ and $\boldsymbol{E_s}$ that are orthogonal and lie in the plane on which $\boldsymbol{k}$ is perpendicular. Moreover, $\boldsymbol{E_p}$ and $\boldsymbol{E_s}$ may have different magnitudes E_{p0} and E_{s0} and different constant phase shifts:

$$\mathbf{E_p} = \mathbf{E_{p0}} \cdot \exp\left(\mathrm{i} \cdot (\mathbf{k} \cdot \mathbf{r} - \omega t + \varphi_p)\right) \tag{4.8}$$

$$\mathbf{E_s} = \mathbf{E_{s0}} \cdot \exp\left(\mathrm{i} \cdot (\mathbf{k} \cdot \mathbf{r} - \omega t + \varphi_s)\right) \tag{4.9}$$

Investigating the locus of $|\mathbf{E}| = \sqrt{E_p^2 + E_s^2}$ in the plane on which $\boldsymbol{k}$ is perpendicular, one can find after some mathematics that E_p and E_s must satisfy the relation

$$\left(\frac{E_p}{E_{p0}}\right)^2 - 2\frac{E_p}{E_{p0}}\frac{E_s}{E_{s0}}\cos\delta + \left(\frac{E_s}{E_{s0}}\right)^2 = \sin^2\delta \tag{4.10}$$

with δ being the difference in the constant phase shifts, $\delta = \phi_p - \phi_s$. Equation (4.10) describes an ellipse in this plane and $\boldsymbol{E}$ rotates along this ellipse. Depending on δ three cases can be distinguished

- linear polarization with $\delta = 0$,
- circular polarization with $\delta = \pi/2$, and
- elliptical polarization with δ being arbitrary, but not 0 and not $\pi/2$.

For linear polarization the field vector does not rotate. In case of circular polarization it is customary to describe the rotation as right-handed if the vector rotates clockwise when viewed in direction opposite to the propagation direction. Accordingly, it is called left-handed if the vector rotates counterclockwise. The three cases of polarization are illustrated in Fig. 4.4.

4.4 Interference

Thomas Young (1773–1829) established in 1801 the principle of *interference* of waves as linear superposition of waves with different phases. The principle is demonstrated in the following for the interference of two waves.

Let $\mathbf{A}(\mathbf{r}, t) = \mathbf{A_0} \cdot \exp(i(\mathbf{k} \cdot \mathbf{r} - \omega t))$ be one wave and $\mathbf{B}(\mathbf{r}, t) = \mathbf{B_0} \cdot \exp(i(\mathbf{k} \cdot \mathbf{r} - \omega t + \phi))$ a second wave with different magnitude having a phase difference ϕ compared to wave $\mathbf{A}$. The linear superposition of $\boldsymbol{A}(\boldsymbol{r}, t)$ and $\boldsymbol{B}(\boldsymbol{r}, t)$ means that a new wave with $\boldsymbol{C}(\boldsymbol{r}, t)$ results from the sum of $\boldsymbol{A}(\boldsymbol{r}, t)$ and $\boldsymbol{B}(\boldsymbol{r}, t)$:

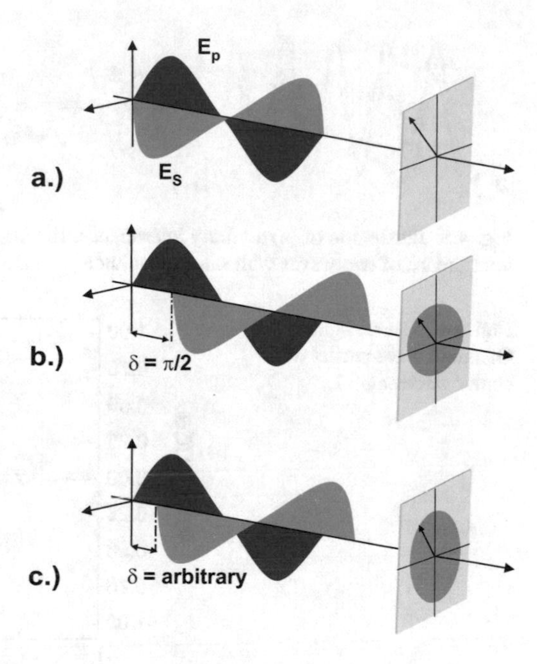

Fig. 4.4 Sketch of (**a**) a linearly polarized wave, (**b**) a circularly polarized wave, and (**c**) an elliptically polarized wave

$$\begin{aligned}\mathbf{C}(\mathbf{r},\mathrm{t}) &= \mathbf{A}(\mathbf{r},\mathrm{t}) + \mathbf{B}(\mathbf{r},\mathrm{t})\\ &= (\mathbf{A_0} - \mathbf{B_0}) \cdot \exp\left(\mathrm{i} \cdot (\mathbf{k} \cdot \mathbf{r} - \omega \mathrm{t})\right) + \mathbf{B_0} \cdot \exp\left(\mathrm{i} \cdot (\mathbf{k} \cdot \mathbf{r} - \omega \mathrm{t})\right) \cdot (1 + \exp(\mathrm{i}\varphi))\\ &= \exp\left(\mathrm{i} \cdot (\mathbf{k} \cdot \mathbf{r} - \omega \mathrm{t})\right) \cdot (\mathbf{A_0} - \mathbf{B_0} + 2\mathbf{B_0} \cdot \exp(\mathrm{i}\varphi/2) \cdot \cos(\varphi/2))\end{aligned} \tag{4.11}$$

The magnitude $C=|\boldsymbol{C}(\boldsymbol{r},t)|$ of the resulting wave strongly depends on the value of the phase difference ϕ: if $\phi=(2m+1)\pi$, with m being an integer number, the magnitude is minimum and even completely vanishes for $\boldsymbol{A_0}=\boldsymbol{B_0}$. On the other hand it becomes maximum for $\phi = 2m\pi$. For identical magnitudes of the two superposed waves complete extinction can be expected, called *destructive interference*, and a doubling of the magnitude, called *constructive interference*. For all other phase differences ϕ an intermediate state occurs.

For illustration, consider two one-dimensional waves with same magnitude at a fixed timepoint t_0. Then, the phase difference is defined by a difference δ in the path along the x-axis and is $\varphi = \frac{2\pi}{\lambda}\delta$. $C(x, t_0)$ follows as

$$\begin{aligned}\mathrm{C}(\mathrm{x},\mathrm{t_0}) &= \mathrm{A} \cdot \sin\left(\frac{2\pi}{\lambda}\mathrm{x} + \omega \mathrm{t_0}\right) + \mathrm{A} \cdot \sin\left(\frac{2\pi}{\lambda}(\mathrm{x}+\delta) + \omega \mathrm{t_0}\right)\\ &= 2\mathrm{A} \cdot \cos\left(\frac{\pi\delta}{\lambda}\right) \cdot \sin\left(\frac{2\pi}{\lambda}\left(\mathrm{x} + \frac{\delta}{2}\right) + \omega \mathrm{t_0}\right)\end{aligned} \tag{4.12}$$

As can be recognized from Fig. 4.5a the superposition of the two waves result in a new wave with magnitude $A<C<2A$ that is phase-shifted compared with the two

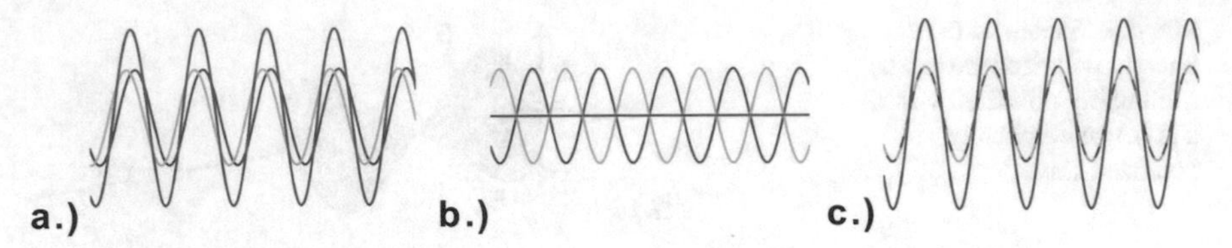

Fig. 4.5 Illustration of (**a**) arbitrary interference, (**b**) destructive interference, and (**c**) constructive interference of two waves with same magnitude

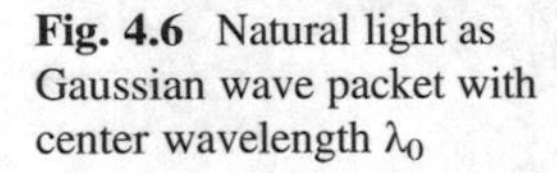

Fig. 4.6 Natural light as Gaussian wave packet with center wavelength λ_0

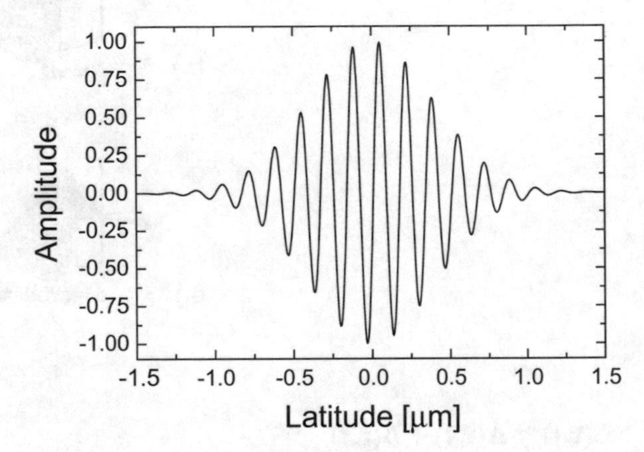

primary waves by $\pi\delta/\lambda$ if the shift δ is arbitrary. For the two special cases $\delta=(2\ m + 1)\lambda/2$ and $\delta = 2m\lambda$ the destructive interference Fig. 4.5b with $C = 0$ and the constructive interference Fig. 4.5c with $C = 2A$ are obtained.

4.5 Coherence

As harmonic waves are extended from minus infinity to infinity linear superposition is possible at each point. This is the ideal case. Looking at a natural light wave the situation is clearly different. The spontaneous emission of light results in a wave packet that can be described by a harmonic wave with center wavelength λ_0 and a Gaussian envelope (see Fig. 4.6). It is the result of the linear superposition of many waves with wavelengths λ around the central wavelength λ_0. It is obvious that two such Gaussian wave packets can only interfere if they do partially overlap. The maximum shift of one wave packet compared to a second wave packet for which interference can still be observed is called *coherence length* l_c. The coherence length depends upon the central wavelength λ_0 in the wave packet and the band width $\Delta\lambda$ of the wave packet

$$l_c = \frac{2 \cdot \ln 2}{\pi} \cdot \frac{\lambda_0^2}{\Delta\lambda} \tag{4.13}$$

Usually $\Delta\lambda$ is the full width at half maximum (FWHM) of the wave packet. For the artificial light source *laser* the bandwidth is extremely small by what a laser has a large coherence length. The smaller the bandwidth $\Delta\lambda$ is the larger is the coherence length of the light source.

4.6 Dielectric Function and Refractive Index

Up to now the electromagnetic wave traveled in free space and in vacuum. Now, the description is extended on the propagation of an electromagnetic wave in matter. The first what can be recognized is that the propagation velocity gets reduced from vacuum light velocity c to $c/n(\lambda)$ caused by the interaction with the matter. This factor is the *refractive index* $n(\lambda)$. Passing through an absorbing medium with *complex refractive index* $n+i\kappa$ also the magnitudes of the electric and magnetic field of the electromagnetic wave get attenuated whereby the *extinction coefficient* or *absorption index* $\kappa(\lambda)$ plays an important role.

From a microscopic view specific electron distributions form around the atoms with increasing number of atoms in the assembly. In a solid state body they are known as *band structure*. Then, also the scope of interactions of light with matter is increased. To describe this interaction, models must be defined that consider the specific electron distribution, the acceleration of the electrons in electric and magnetic fields, and exchange and interaction effects among the electrons. The most precise description is only possible using quantum mechanics but also from classical mechanics very well suited descriptions are available. The result of all these examinations is the *complex dielectric function* $\varepsilon(\lambda)=\varepsilon_1(\lambda)+i\cdot\varepsilon_2(\lambda)$. With the dielectric function a quantity is introduced that gives information on magnitude and type of interaction of the electromagnetic wave with matter. Macroscopically, the real part $\varepsilon_1(\lambda)$ is related to the polarizability of the matter while the imaginary part $\varepsilon_2(\lambda)$ is a measure for the absorption. Hence, the dielectric function is the more proper quantity to describe the interaction of electromagnetic radiation with matter while the refractive index is more proper for the description of the propagation of the wave in matter. Dielectric function and refractive index are connected by Maxwell's relation

$$n + i\kappa = \sqrt{\varepsilon_1 + i\varepsilon_2} \tag{4.14}$$

This relation can be rewritten either for the dielectric function as

$$\varepsilon_1 = n^2 - \kappa^2, \tag{4.15}$$

$$\varepsilon_2 = 2 \cdot n \cdot \kappa, \tag{4.16}$$

or for the refractive index as

$$n = \sqrt{\frac{\varepsilon_1}{2} + \frac{1}{2}\sqrt{\varepsilon_1^2 + \varepsilon_2^2}}, \tag{4.17}$$

$$\kappa = \sqrt{-\frac{\varepsilon_1}{2} + \frac{1}{2}\sqrt{\varepsilon_1^2 + \varepsilon_2^2}}. \tag{4.18}$$

The dielectric function exhibits a particularity: the *anisotropy* of the dielectric function. This pecularity is not obvious. However, when going into the details of the electronic structure of a crystal one can find a dependence of the electron distribution on the crystalline structure. The distribution of electrons is not equal in each crystalline direction. This is relevant mainly for trigonal, tetragonal, and orthorhombic crystal structures. In consequence, the dielectric function is also different in different directions because the induced dipoles differ for each crystalline direction. This anisotropy results in an *ordinary ray* that obeys the common rules in optics (reflection law, Snell's law) and in at least one *extraordinary ray* with specific description and rules. The dielectric function is now a tensor of second rank. The consequential variation of the refractive index with direction of incident light can be represented by an ellipsoid called *indicatrix*. For cubic crystals the indicatrix is a sphere. For uniaxial crystals it has a circular symmetry around the optical axis and for biaxial crystals it is a general ellipsoid with only twofold symmetry.

As mentioned above the complex dielectric function can be derived from classical mechanics when considering the movement and acceleration of electrons in electric and magnetic time-periodic fields of an electromagnetic wave. This was done by H. A. Lorentz in 1895 [1]. The result is the *harmonic oscillator model* or *Lorentz oscillator* of the dielectric function

$$\varepsilon(\omega) = 1 + \sum_n \frac{S_n \omega_n^2}{\omega_n^2 - \omega^2 - i\omega\gamma_n} \tag{4.19}$$

with S_n being the oscillator strength, ω_n is the resonance frequency, and γ_n the damping constant of the n-th harmonic oscillator. Figure 4.7 exemplarily shows the dielectric function and the corresponding refractive index of a harmonic oscillator with resonance frequency $3.5 \cdot 10^{15}$ s^{-1}, damping constant $3.5 \cdot 10^{14}$ s^{-1}, and oscillator strength $S = 1$.

At the resonance frequency the imaginary part exhibits a maximum and rapidly decreases to the right and the left of the resonance frequency. Far from the resonance frequency ε_2 vanishes. The real part decreases at high frequencies (low wavelengths) when approaching the resonance frequency and even becomes negative. In the vicinity of the resonance frequency the real part changes very rapidly to high positive

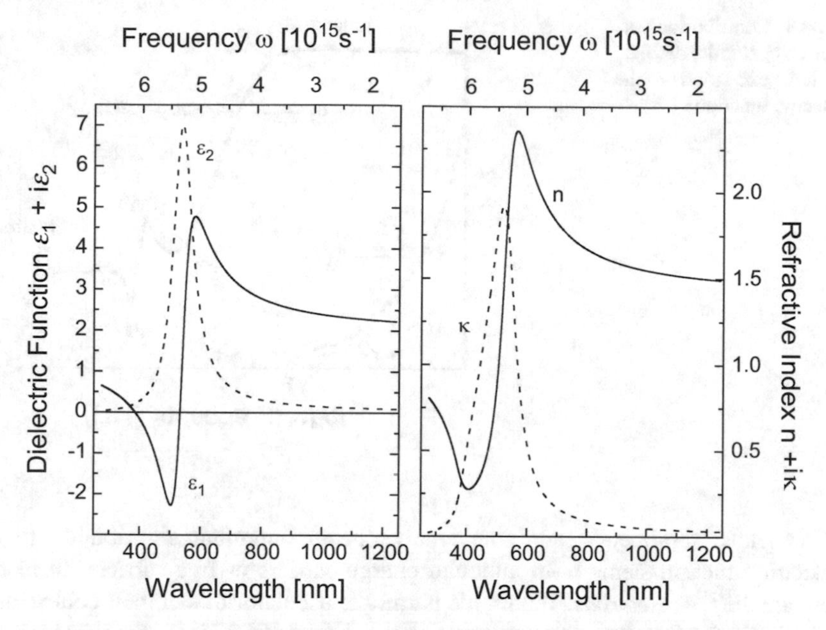

Fig. 4.7 Dielectric function and refractive index of a harmonic oscillator with oscillator strength $S = 1$, resonance frequency $3.5{\cdot}10^{15}\ \mathrm{s}^{-1}$, and damping constant $3.5{\cdot}10^{14}\ \mathrm{s}^{-1}$

values from which it continuously decreases again when going to low frequencies. The decrease with decreasing frequency (increasing wavelength) is called *normal dispersion* while in the region of rapid increase with decreasing frequency it is called *anomal dispersion*.

In general, the harmonic oscillator model can be applied in all cases where the force fields of the electromagnetic wave can induce electric dipoles by separating the center of gravity of positive charges and negative charges. Therefore, not only electronic excitations can be described with this model but also atomic excitations in bipolar crystals like NaCl or excitations of dipoles like in water. Then, the dielectric function is the sum over all the susceptibilities resulting from these excitations. As the ions in bipolar crystals are much heavier than electrons the corresponding atomic excitations lie at low frequencies in the far infrared where the negatively and positively charged ions are displaced in different directions. At frequencies in between all the excitations of dipoles due to a change of the orientation of the dipoles caused by the electric field can be found (see Fig. 4.8).

The harmonic oscillator model cannot always be applied. The biggest restriction results from the assumption of a linear response to the applied electric field. For large electric fields the response of condensed matter on these large electric fields becomes dependent on higher powers of the fields. This phenomenon is called *optical nonlinearity* in contrast to the linear response at moderate intensities. Optical nonlinearities are not additive.

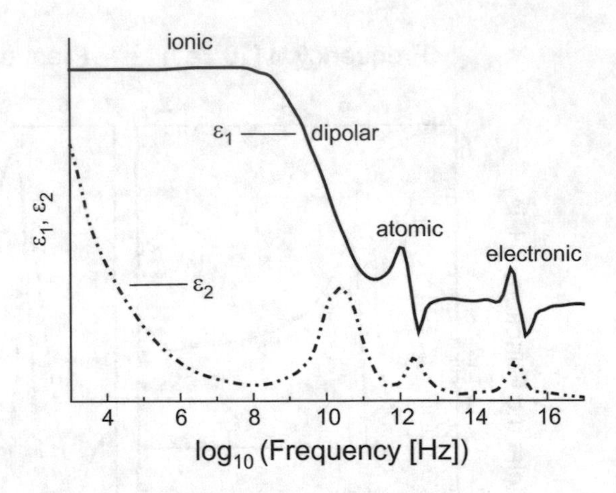

Fig. 4.8 Contributions of electronic, atomic, dipolar, and ionic excitations to the dielectric function

In metals, semimetals, and semiconductors an important contribution to the dielectric function stems from unbound charge carriers or free carriers. In metals these are the *free electrons*. Within the harmonic oscillator model their contribution is obtained when assuming the resonance frequency $\omega_{fc} = 0$. Then, the susceptibility of the unbound charge carriers – the *Drude susceptibility* [2, 3] – reads:

$$\chi_{fc}(\omega) = -\frac{\omega_P^2}{\omega^2 + i\omega\gamma_{fc}} \tag{4.20}$$

with the abbreviation

$$\omega_P^2 = \frac{Ne_0^2}{V\varepsilon_0 m_{eff}} \tag{4.21}$$

being the *plasma frequency of the unbound charge carriers* assuming them as a plasma. In a parabolic band structure the effective mass m_{eff} corresponds to the electron mass m_e but in nonparabolic band structures m_{eff} may differ from m_e. For example, in silicon the negative free carriers in n-doped silicon have an effective mass of $m_{eff} = 0.8 \cdot m_e$ while the positive carriers in p-doped silicon have an effective mass of $m_{eff} = 0.26 \cdot m_e$.

The harmonic oscillator model is quite often a good choice for the description of the dielectric function. But sometimes mainly material properties give reason for an unsufficient description of the dielectric function with the harmonic oscillator model. For these cases several extensions were developed that take care of specific material properties. They are summarized below.

For statistically pertubated or amorphous materials extensions of the harmonic oscillator model exist as *Brendel oscillator* [4] or as *Kim oscillator* [5]. Both are harmonic oscillators but with a width γ_n that is inhomogeneously broadened by an infinite sum over sharp harmonic oscillators with eigenfrequency ω_x and width γ_n (Brendel) or is assumed to be frequency dependent (Kim).

Amorphous semiconductor and oxide materials often have optical functions that depend upon deposition conditions. Their optical constants also cannot be described by an unmodified harmonic oscillator. A first approach for the modified imaginary part of the dielectric function of these materials stems from Tauc et al. [6]. Jellison and Modine [7] derived a model based on a combination of the Tauc band egde and the Lorentz oscillator formulation, the *Tauc-Lorentz model.*

In the Tauc-Lorentz model the imaginary part of the complex dielectric function of amorphous materials with band gap (mainly semiconductor materials) can be modeled as

$$\varepsilon_{2,\mathrm{TL}}(\omega) = \begin{cases} \dfrac{\left(\omega - \omega_{\mathrm{gap}}\right)^2}{\omega^2} \cdot \dfrac{S \cdot \omega_0^2 \cdot \gamma \cdot \omega}{\left(\omega^2 - \omega_0^2\right)^2 + \omega^2\gamma^2} & \omega > \omega_{\mathrm{gap}} \\ 0 & \omega \le \omega_{\mathrm{gap}} \end{cases} \tag{4.22}$$

The oscillator has the resonance frequency ω_0 and a damping constant γ. ω_{gap} is the frequency corresponding to the band gap energy $E_{gap} = \hbar\omega_{gap}$. The real part $\varepsilon_{1,TL}$ is obtained from the imaginary part using Kramers-Kronig relations [8, 9] sometimes called dispersion integrals.

An extension of the Tauc-Lorentz model is the *Cody-Lorentz model* from Ferlauto et al. [10]. A further model for amorphous semiconductors is well known as *OJL-model* from O'Leary, Johnson and Lim [11]. In all cases a modified imaginary part ε_2 of the dielectric function results and the real part ε_1 must be calculated using Kramers Kronig relations.

Optical constants n+iκ or ε_1+iε_2 can always be modeled using the physical models presented above. However, in the past two centuries empiric formulas for the refractive index $n(\lambda)$ were developed and are still in use to parametrize the refractive index. One of the most prominent formula is the *Sellmeier formula* [12].

$$n^2 - 1 = \sum_{j=1}^{N} \frac{A_j\lambda^2}{\lambda^2 - B_j}. \tag{4.23}$$

This formula is the most physical since it corresponds to the sum over N undamped harmonic oscillators with eigenfrequencies $\omega_j = \frac{2\pi c}{\sqrt{B_j}}$ and oscillator strengths S_j=A_j. The most important modification of this formula is to replace n^2-1 by $n^2 - n_0^2$.

Another often used formula is the *Schott formula*. Originally developed by Erich Schott at SCHOTT AG in 1966 and also used by SCHOTT AG until 1992, it is nowadays used from other glass manufacturers like CORNING Inc., HOYA Inc., HIKARI Inc., or SUMITA Inc. The general form is

$$n^2 = \sum_{j=0}^{N} A_j \lambda^{2j} + \sum_{k=1}^{M} B_k \lambda^{-2k}. \qquad (4.24)$$

The original Schott formula is obtained for $N = 1$ and $M = 4$. Eq. (4.24) is also known as *Laurent formula*, because it corresponds to a Laurent series in the wavelength λ.

The third important and often used empiric formula is the *Cauchy formula* from the prominent mathematician A. L. Cauchy [13, 14]:

$$n = A_n + \frac{B_n}{\lambda^2} + \frac{C_n}{\lambda^4} \qquad \kappa = A_\kappa + \frac{B_\kappa}{\lambda^2} + \frac{C_\kappa}{\lambda^4}. \qquad (4.25)$$

The advantage of the Cauchy formula compared to Sellmeier and Schott formula is that it also considers the imaginary part κ of the complex refractive index. It is therefore suited to fit also the optical constants of absorbing materials. It is often used for photoresists which are absorbing in the UV and at wavelengths in the violet/blue visible spectral region.

In close relation to the Cauchy formula another formula was developed, the *exponential Cauchy formula*:

$$n = A_n + \frac{B_n}{\lambda^2} + \frac{C_n}{\lambda^4} \qquad \kappa = A_\kappa \exp\left(B_\kappa \left(\frac{1.239841875}{\lambda} \right) - C_\kappa \right). \qquad (4.26)$$

The difference to the Cauchy formula is in the ansatz for the absorption index κ which now is described by an exponential function. The value 1.239841875 is valid for wavelengths in μm.

In fact, many optical constants can be approximated by these few empiric formulas. Besides these formulas a series of more or less known empiric formulas exist that are applied to transparent materials. For more details see e.g. [15]. and the references therein.

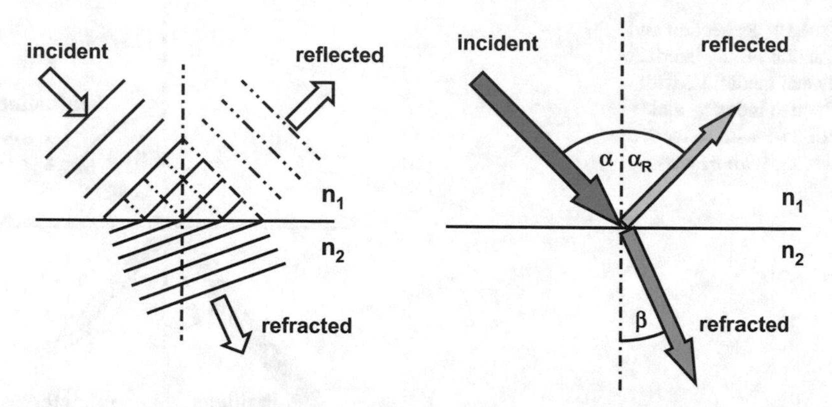

Fig. 4.9 Reflection and refraction on the interface between medium 1 with refractive index n_1 and medium 2 with refractive index n_2, with $n_1 < n_2$

4.7 Reflection and Refraction

After having discussed in Sect. 4.6 how an electromagnetic wave travels in a medium with refractive index, now a plane interface as boundary between two different materials is introduced and the resulting wave propagation in each of the two materials is considered. The interface to the second medium strongly disturbs the free propagation of the wave in the first medium whereby new waves are obtained by reflection and refraction.

Consider a plane interface between two media 1 and 2 with refractive indices n_1 and n_2. For the moment n_2 is assumed to be real valued, i.e. medium 2 is nonabsorbing. Medium 1 is always assumed to be nonabsorbing in the following. A plane wave in medium 1 with wave vector $\boldsymbol{k_{inc}}=k_{inc}\cdot(\sin\alpha,-\cos\alpha,0)^T$ hits this interface so that its propagation direction encloses an angle α with the normal to the interface. Applying now Huygens' principle the wavefronts in medium 1 and medium 2 can be constructed. This is illustrated in Fig. 4.9.

When constructing the wavefronts in medium 1, a second wave with wave vector $\boldsymbol{k_{ref}}=k_{inc}\cdot(\sin\alpha,\cos\alpha,0)^T$ is obtained. It is *reflected* at the interface between the two media. It also includes the angle α to the normal on the interface. From this the *reflection law* can be deduced.

$$\text{Angle of Reflection } \alpha_R = \text{Angle of Incidence } \alpha \tag{4.27}$$

which is well-known since the ancient (Euclid, 300 B.C.).

When constructing the wavefronts in medium 2, it must be taken into account that the wavelength changes from λ/n_1 to λ/n_2. If $n_2 > n_1$ the resulting wave with $\boldsymbol{k_{refr}}$ in medium 2 encloses an angle β with the normal that is smaller than α because the distance of planes of constant phase – the wavelength – becomes smaller than in medium 1. The relation between β and α is not at all as obvious as the relation for α_R

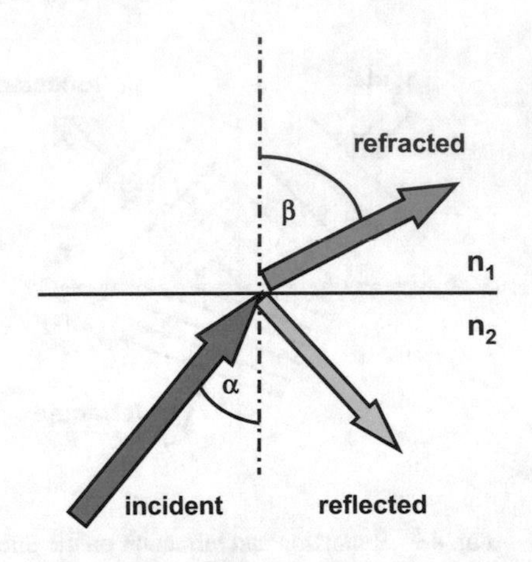

Fig. 4.10 Reflection and refraction on the interface between medium 2 with refractive index n_2 and medium 1 with refractive index n_1, with $n_1 < n_2$

and α in the reflection law. Already Ptolemy (approx. 90–168AD.) studied the relation between α and β but did not arrive in a formula. It was the Dutch Willebrord van Roijen Snell who derived first a mathematical relation in 1618 which is well-known as *Snell's law of refraction*

$$n_1(\lambda) \cdot \sin(\alpha) = n_2(\lambda) \cdot \sin(\beta) \quad (4.28)$$

Snell died before he could publish his result but he reported his French scientific colleague René Descartes. Descartes published the refraction law later in 1637 in his *Discours de la méthode pour bien conduire sa raison et chercher la verité dans les sciences*.

Consider now the reverse situation: the wave is coming from the medium with higher refractive index n_2 (optically thick) and is going to the medium with lower refractive index n_1 (optically thin) (see Fig. 4.10). From Snell's law (Eq. (4.28)) it follows that for an angle of incidence $\alpha_C = \sin^{-1}\left(\frac{n_1}{n_2}\right)$ – this is called *critical angle of total reflection* - the maximum angle of refraction of $\beta = 90°$ is reached. The corresponding refracted beam then travels along with the interface in the optically thinner medium. For $\alpha > \alpha_C$ the incident light gets totally reflected at the interface. In the optically thinner medium there still exists a wave propagating along with the interface but with the amplitude of the electromagnetic fields decreasing exponentially with $\exp(-\gamma z)$ in the direction perpendicular to the interface.

This kind of waves is called *evanescent wave*. The attenuation constant γ depends upon the angle of incidence α and the refractive indices n_1 and n_2:

$$\gamma = \frac{2\pi}{\lambda}\left(n_2^2 \sin^2\alpha - n_1^2\right)^{1/2}. \quad (4.29)$$

For $\alpha = \alpha_C$ it is $\gamma = 0$ and for $0 \leq \alpha < \alpha_C$ it becomes purely imaginary, so that the wave is not attenuated in z-direction but corresponds to a radiating wave, as expected from Snell's law.

Up to this point only the propagation direction of the waves has been considered. The magnitudes of the reflected and refracted electromagnetic fields are related to the magnitude of the incident wave. The relations -called *Fresnel equations*- follow from applying Maxwell's boundary conditions at the surface. Passing on the derivation here only the results are given as *Fresnel coefficients*. Note that they depend upon the polarization of the incident wave and are different for s- and p-polarization:

$$r_s = \frac{E_{s,\,ref}}{E_{s,\,inc}} = \frac{n_1 \cos(\alpha) - n_2 \cos(\beta)}{n_1 \cos(\alpha) + n_2 \cos(\beta)}, \tag{4.30}$$

$$t_s = \frac{E_{s,\,refr}}{E_{s,\,inc}} = \frac{2n_1 \cos(\alpha)}{n_1 \cos(\alpha) + n_2 \cos(\beta)}, \tag{4.31}$$

$$r_p = \frac{E_{p,\,ref}}{E_{p,\,inc}} = -\frac{n_1 \cos(\beta) - n_2 \cos(\alpha)}{n_1 \cos(\beta) + n_2 \cos(\alpha)}, \tag{4.32}$$

$$t_p = \frac{E_{p,\,refr}}{E_{p,\,inc}} = \frac{2n_1 \cos(\alpha)}{n_1 \cos(\beta) + n_2 \cos(\alpha)}. \tag{4.33}$$

The corresponding intensities R_s, T_s, R_p, and T_p are obtained by considering the incident, reflected, and refracted energy flux density. It follows that

$$R_{s,\,p} = r_{s,\,p} \cdot r^*_{s,\,p} \tag{4.34}$$

$$T_{s,\,p} = \frac{n_2 \cos(\beta)}{n_1 \cos(\alpha)} \cdot t_{s,\,p} \cdot t^*_{s,\,p} \tag{4.35}$$

where the asterisk denotes the complex conjugate (for numerics with complex numbers see the Appendix).

What happens with the reflection at the surface S and the transmission through the surface including Snell's law if the medium 2 becomes absorbing, i.e. the refractive index of medium 2 becomes complex valued? The first obvious consequence is that sin(β) in Eq. (4.28) and hence the angle of refraction β must also be complex. It follows that

$$\mathrm{Re}\,(\sin(\beta)) = \frac{n_2}{n_2^2 + \kappa_2^2} n_1 \sin(\alpha), \tag{4.36}$$

$$\mathrm{Im}(\sin(\beta)) = -\frac{\kappa_2}{n_2^2 + \kappa_2^2} n_1 \sin(\alpha). \tag{4.37}$$

and the Fresnel coefficients r_s, t_s, r_p, and t_p become complex numbers. Then, the planes of equal phase and the planes of equal magnitude enclose an angle and do not further coincide for the propagating wave in the absorbing medium.

4.8 Dispersion Effects

Already Sect. 4.6 showed that the dielectric function or the refractive index are not constant but are wavelength-dependent. As an example the wavelength dependence of the refractive index of two optical glasses manufactured by SCHOTT AG, Germany, namely the glasses N-BK7 and SF6, is depicted in Fig. 4.11.

As can clearly be recognized the refractive index decreases with increasing wavelength for both glasses. This *dispersion* of the optical constants strongly affects reflection and refraction. Blue light (short wavelengths) gets stronger refracted than red light (long wavelengths) as the refractive index is higher for blue light. Also the critical angle of total reflection is different for blue light and red light. This dispersion effect is mostly unfavorable for optical components with curved surfaces.

What happens at a curved surface when a beam of parallel light hits the surface? This is illustrated in the next Fig. 4.12.

For a transparent medium with convex curvature the reflection law yields a divergence of the incident parallel beam and Snell's law a focusing of the beam at a certain point on the optical axis. The locus of this focal point depends on the radius of curvature *R* and the refractive index. For a concave curvature the reflected beams

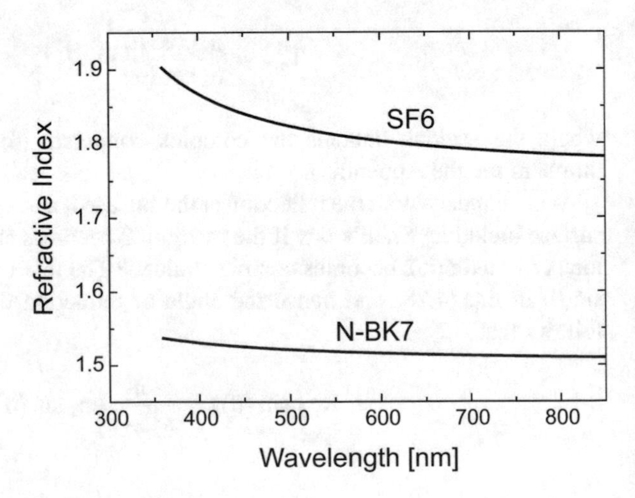

Fig. 4.11 Wavelength-dependence of the refractive index of the two optical glasses N-BK7 and SF6 from SCHOTT AG, Germany

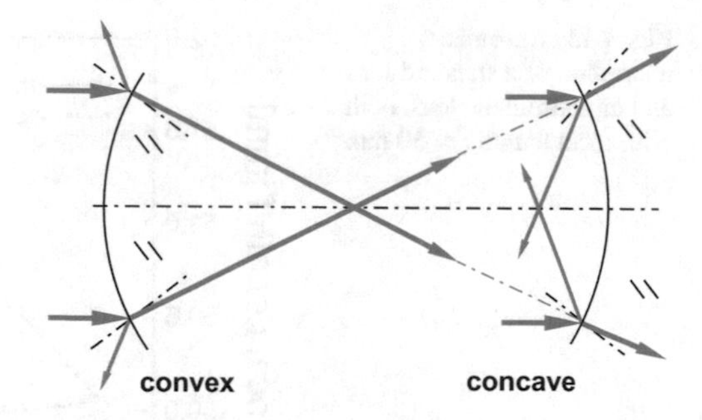

Fig. 4.12 Reflection and refraction on curved surfaces

get focused in a point in front of the curved surface while the refracted beams diverge. The refracted beams apparently come from a virtually focal point on the optical axis in front of the surface. For a spherically shaped mirror with radius R only reflected beams are obtained. That means for a convex curvature a divergence of the beam and for a concave curvature a focus that is determined only by the radius of curvature. The focal length is then $f = R/2$ for the concave mirror and $f = -R/2$ for the convex mirror.

Convex, concave, and plane surfaces can be combined to form lenses which are either focusing or diverging. In doing so a plane surface is described by an infinitely large radius of curvature. In addition, the distance d measured on the optical axis between the front and the rear surface plays an important role for the resulting focal length f or its reciprocal value, the *refractive power* $D = 1/f$. For a combination of two curved surfaces with radii of curvature R_1 and R_2 and distance d the refractive power is

$$D = \frac{1}{f} = (n(\lambda) - 1) \cdot \left(\frac{1}{R_1} - \frac{1}{R_2} + \frac{(n(\lambda) - 1) \cdot d}{n(\lambda) \cdot R_1 \cdot R_2} \right) \tag{4.38}$$

The focal length depends on the wavelength due to the dispersion of the refractive index. Therefore, when illuminating a focusing lens with a parallel white light beam, the focal length is different for different colors respectively wavelengths. The higher refractive index for blue light causes stronger refraction as for red light. Then, also the focal point lies closer to the lens for blue light than for the red light. This *chromatic aberration* must be avoided when using lenses in optical imaging. It can be strongly reduced by combining a focusing lens of a low refractive crown glass with a dispersing lens of high refractive flint glass. The focal length of this achromatic lens is almost constant over a wide spectral range as can be seen from Fig. 4.13 for a standard lens and an achromatic lens both with focal length $f = 50$ mm.

When using mirrors as imaging elements the chromatic aberration is absent. Spherical aberrations and distortions must however be corrected just as they must be corrected for lenses (aspheric shapes, gradients).

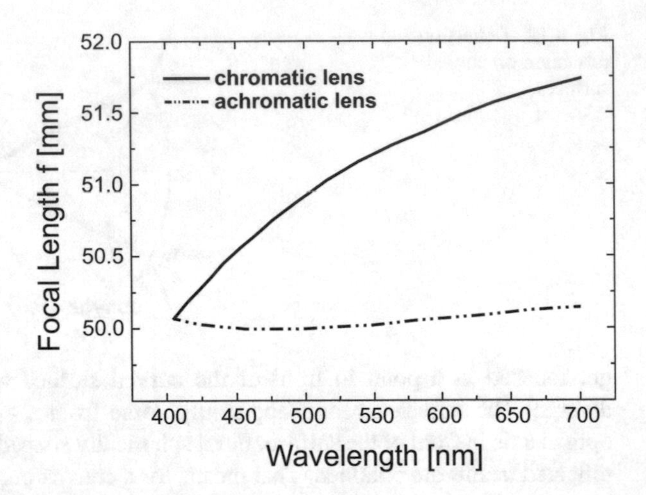

Fig. 4.13 Chromatic aberration of a standard lens and an achromatic lens, both with focal length $f = 50$ mm

4.9 Diffraction

Unless for reflection and refraction, the change in the propagation direction in *diffraction* is accompanied by the linear superposition of partial waves from different sites that have the same propagation direction but a different phase.

The two simplest cases where diffraction can be obtained are a single slit in an opaque diaphragm or a pinhole in an opaque diaphragm. For the slit the energy of the wave in the slit is distributed in the half space behind the slit and can be described by a squared sinc-function $\text{sinc}^2(\beta)$:

$$J_{\text{slit}}(\beta) \propto b^2 \frac{\sin^2\left(\frac{\pi b}{\lambda} \sin\beta\right)}{\left(\frac{\pi b}{\lambda} \sin\beta\right)^2} \tag{4.39}$$

where β is the diffraction angle and b is the width of the slit.

Only the minima of this energy distribution always positive definite function are directly obvious. They are obtained in the directions where $\sin(\beta)$ fulfills

$$\sin\beta_m = m\frac{\lambda}{b}, \qquad m = \pm 1, \pm 2, \pm 3, \ldots \tag{4.40}$$

For the maxima the condition is more complex

$$\tan\left(\frac{\pi b}{\lambda} \sin\beta\right) = \frac{\pi b}{\lambda} \sin\beta. \tag{4.41}$$

Commonly, the resolution of a single slit is insufficient for use in metrological applications. A drastic improvement is obtained when using a *grating*. A grating consists of a thin plane plate with periodically arranged slits illuminated by a plane

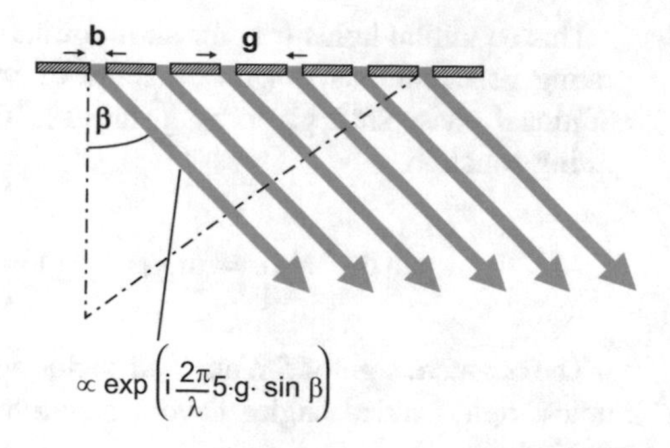

Fig. 4.14 Sketch of a grating with partial waves diffracted in the direction given by the diffraction angle β

wave. The periodicity of the slits is the *grating period g*. The reciprocal of the grating period is called *groove density* $L = 1/g$ (Fig. 4.14).

From each slit a wave propagates in the direction given by the diffraction angle β. From slit to slit they have the phase difference $\phi = \frac{2\pi}{\lambda} g \cdot \sin\beta$ and a phase factor $\exp(i\varphi)$. These waves superpose in the far field (far from the grating surface). Then, the radiant flux through the slits of the grating into the half space behind the grating is

$$
\begin{aligned}
J_{grating}(\beta) &\propto J_{slit}(\beta) \cdot \left| \sum_{n=0}^{N-1} (\exp(i\phi))^n \right|^2 \\
&\propto \frac{\sin^2\left(\frac{\pi b}{\lambda}\sin\beta\right)}{\left(\frac{\pi b}{\lambda}\sin\beta\right)^2} \cdot \frac{\sin^2\left(\frac{\pi N g}{\lambda}\sin\beta\right)}{\sin^2\left(\frac{\pi g}{\lambda}\sin\beta\right)}.
\end{aligned}
\tag{4.42}
$$

The intensity diffracted by the grating into direction of the angle β consists of two factors: the first $J_{slit}(\beta)$ describes the diffraction by a single slit and the second results from the multiple interference of the waves coming from all *N* slits of the grating. This second factor determines the diffracted light in a distinct manner as it introduces additional minima and maxima. If the single slit diffraction leads to zero intensity in direction of the angle β, this minimum is kept in the intensity diffracted by the grating. However, between two minima of $J_{slit}(\beta)$ the second factor -the *grating function* GF- can be zero with maxima of GF in between. A further effect of the multiple interference is that the intensity of these new maxima from the GF strongly increases with the number of illuminated slits *N* proportional to N^2 so that these maxima dominate the intensity distribution in the half space behind the grating. This is a consequence of the energy conservation law.

The maxima of GF are found after some mathematics. The first order derivation of GF yields the conditions for extrema and the second order derivative finally shows which of the extrema are the maxima. The prominent maxima are given by

$$
\sin\beta = m\frac{\lambda}{g}, \qquad m = 0, \ \pm 1, \pm 2, \pm 3, \ldots \tag{4.43}
$$

where *m* is the order of diffraction.

This condition holds true for an illumination perpendicular to the grating. If the grating gets illuminated with an angle of incidence ε the partial waves have an additional phase shift given by $g \cdot \sin(-\varepsilon)$. Then, the maxima of the GF obey the grating equation

$$\sin\beta - \sin\varepsilon = m\frac{\lambda}{g}, \qquad m = 0, \ \pm 1, \pm 2, \pm 3, \ldots \tag{4.44}$$

The negative sign of the angle of incidence results from the sign convention for angles: right-handed angles have a negative sign, and left-handed angles have a positive sign.

Please note that for a *reflection grating*, the sign of the angle of incidence ε and the sign of the angle of diffraction β are identical by convention. Therefore, for a reflection grating the grating equation reads

$$\sin\beta + \sin\varepsilon = m\frac{\lambda}{g}, \qquad m = 0, \ \pm 1, \pm 2, \pm 3, \ldots \tag{4.45}$$

4.10 Scattering

Reflection, refraction, and diffraction of light and other electromagnetic radiation represent a redirection of the light from its original path into distinct directions given by the reflection law, Snell's law, or the grating equation.

Unlike these processes *scattering* of radiation distributes the light in all directions but with different magnitudes in different angles. Thereby the ratio (size of the scatterer)/wavelength plays an important role for the spatial distribution of light. A scatterer very small compared to the wavelength (Rayleigh limit) scatters the light symmetrically in forward and backward direction. With increasing size forward scattering dominates the spatial distribution. Not only the size of the scatterer affects light scattering but also its shape, constitution, and refractive index. Finally, the concentration of scatterers in a volume influences the spatial distribution of scattered light. Densely packed scatterers lead to an almost homogeneous distribution of scattered light like for example in clouds, fog, dispersion colors, or paper. In each case the scattering diminishes the radiation flux in propagation direction of the reflected, refracted, or diffracted light by the redistribution of the incident light into all solid angles.

For this book the scattering caused by surface imperfections is of interest as opposed to scattering from individual molecules and particles as scatterer. The latter can be read more detailed in several monographs [16–21]. Scattering at surfaces is comprehensively discussed in the monographs [22–24]. Surface imperfections may by scratches, recesses, or embossments. A widespread distribution of such embossments and recesses on a surface leads to the surface roughness. Light scattering

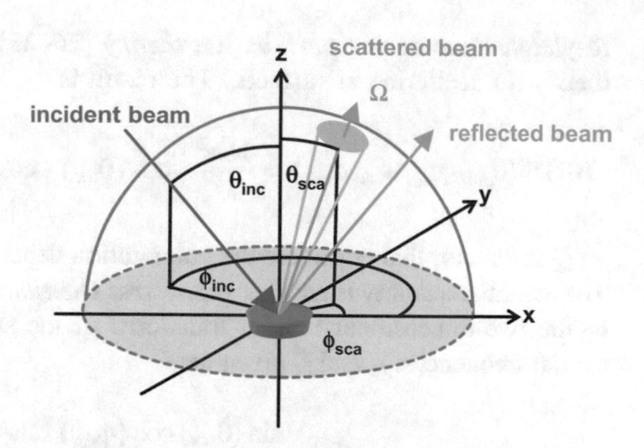

Fig. 4.15 Defining the geometry of the BRDF

belongs to area integrating methods for measuring surface texture. The used methods probe an area of the surface altogether and yield parameters that are characteristical for the texture of the area as a whole.

The *bidirectional scatter distribution function* (BSDF) is commonly used to describe the intensity distribution of scattered light. The terms BRDF, BTDF, and BVDF used for reflective, transmissive, and volume functions are merely subsets of the more-generic BSDF. In applications the most often used subset is the *bidirectional reflection distribution function* BRDF. Its derivation and notation goes back to Nicodemus et al. [25] who examined extensively the problem of measuring and defining the reflectance of optics that is partially scattering. The defining geometry is shown in Fig. 4.15.

For a collimated beam which is incident at angle θ_{inc} on a surface a part of the incident light gets scattered into the solid angle Ω having an angle θ_{sca} to the normal on the surface and an azimuth ϕ_{sca}. In terms of radiometric quantities the BRDF is then defined as the surface radiance divided by the incident surface irradiance. The surface irradiance P_i is the light flux incident on the illuminated surface area and the scattered surface radiance is the light flux P_{sca} scattered through the solid angle Ω normalized to the projected solid angle. The projected solid angle is the solid angle multiplied by $\cos(\theta_{sca})$. From this the BRDF is

$$\mathrm{BRDF}(\theta_{inc}, \phi_{inc}, \theta_{sca}, \phi_{sca}) \approx \frac{P_{sca}}{P_{inc} \cdot (\Omega \cdot \cos(\theta_{sca}))}. \tag{4.46}$$

Angle resolved scattering ARS is identical to the BRDF scaled by a factor cos (θ_{sca}):

$$\mathrm{ARS} = \mathrm{BRDF} \cdot \cos(\theta_{sca}). \tag{4.47}$$

With the approximation that the roughness of a surface is only a small pertubation from the ideally plane surface, i. e. for smooth surfaces, the BRDF can be calculated from the surface topograpy in the so-called *Rayleigh-Rice* or

Rayleigh-Fano vector pertubation theory [26–28]. It is the oldest approximate theory for scattering at surfaces. The result is

$$\mathrm{BRDF}(\theta_{inc}, \phi_{inc}, \theta_{sca}, \phi_{sca}) = \frac{16\pi^2}{\lambda^4} \cdot \cos(\theta_{inc}) \cdot \cos(\theta_{sca}) \cdot Q \cdot \mathrm{PSD}(f_x, f_y) \quad (4.48)$$

Q is a factor that considers the polarization dependent reflectance of the surface. The essential quantity is the *Power Spectral Distribution* function PSD. It is identical to the two-dimensional Fourier transform of the surface topography $z(x, y)$ with spatial frequencies f_x and f_y given as

$$f_x = \frac{\sin(\theta_{sca})\cos(\phi_{sca})}{\lambda} - \frac{\sin(\theta_{inc})}{\lambda}, \quad (4.49)$$

$$f_y = \frac{\sin(\theta_{sca})\sin(\phi_{sca})}{\lambda} \quad (4.50)$$

The Rayleigh-Rice theory is not readily applicable to rough surfaces since is valid only for smooth surfaces with r.m.s. roughness parameters R_q that fulfill the condition

$$\left(\frac{4\pi R_q \cdot \cos(\theta_{inc})}{\lambda}\right)^2 \ll 1 \quad (4.51)$$

For this reason, Beckmann [22] and Bruce and Dainty [29–31] (as well as others) used the Kirchhoff method to approach the scattering problem. Instead of satisfying the exact boundary conditions, the field and its normal derivative are approximated on the scattering surface. Although also this approximation causes the loss of some generality, it allows to calculate scattering for surfaces much rougher than the Rayleigh-Rice theory allows.

In 1976 Harvey and Shack [32] formulated a scattering theory in a linear systems format resulting in a surface transfer function that relates scattering behavior to surface topography. Yet, the original Harvey-Shack theory had limitations and needed improvements. They were introduced 2006 in the Generalized Harvey-Shack Theory [33]. In this theory the BRDF follows from the transfer function as

$$\mathrm{BRDF}(\theta_{inc}, \phi_{inc}, \theta_{sca}, \phi_{sca}) = \frac{4\pi^2}{\lambda^4} \cdot (\cos(\theta_{inc}) + \cos(\theta_{sca}))^2 \cdot Q \cdot \mathrm{PSD}(f_x, f_y) \quad (4.52)$$

When integrating the BRDF over the full halfspace the totally scattered flux P_{sca} is obtained. The *total integrated scattering* (TIS) is defined by relating this flux to the sum of scattered and regular reflected flux P_{refl}

$$\text{TIS} = \frac{P_{sca}}{P_{refl} + P_{sca}}. \tag{4.53}$$

For perfectly conductive surfaces (reflectivity $R_O = 1$) with a Gaussian height distribution Davies [34] related TIS to the r.m.s. surface roughess parameter R_q by

$$\text{TIS} \cong \left(\frac{4\pi R_q}{\lambda}\right)^2. \tag{4.54}$$

From this relation follows that surfaces scatter more at shorter wavelengths than at longer wavelengths as it is also the finding in scattering by particles. It was shown by Church et al. [35] that the assumption of a Gaussian height distribution is not necessary to obtain this relation. On the basis of Davies' result Bennett and Porteus [36] found a functional relationship between TIS and surface roughness

$$\text{TIS} = R_0 \cdot \left(1 - \exp\left(-\left(\frac{4\pi R_q \cos\theta_{inc}}{\lambda}\right)^2\right)\right). \tag{4.55}$$

In this equation R_O is the theoretical reflectance of the surface. This equation was later extended also to refraction. What can be learned from this relation in addition to Eq. (4.54) is that optical scattering is proportional to reflectance. This means that surfaces intended to reflect light will inherently scatter more light than transmissive surfaces. Second, more light gets scattered at normal incidence than at grazing incidence. These results correlate well with experience and observation. For r.m.s. roughness values much smaller than the wavelength the exponential in Eq. (4.55) can be approximated by the first and second term in its series expansion from which a simpler relation for TIS is obtained assuming $\theta_{inc} = 0°$ [37].

$$\text{TIS} = R_0 \cdot \left(\frac{4\pi R_q}{\lambda}\right)^2. \tag{4.56}$$

This result was confirmed by Harvey et al. [38]. It corresponds to Davies' result for $R_O = 1$.

Without describing mathematical details, ISO 10110-8 [39] characterizes the close relationship between the surface roughness and the light scattering properties of optical surfaces and describes drawing indications that one can use to specify r.m.s. roughness, r.m.s. slope, and power spectral density for an optical surface. As light scattering is also often used for characterization of semiconductors corresponding standards are formulated in the semiconductor industry with the SEMI MF 1048–1109 and SEMI ME 1392–1109 norms [40, 41]. Note that the r.m.s. roughness (R_q or S_q) calculated from a model using scattered light measurements is not identical to the R_q or S_q value calculated from a profile or a topography image. This is principally due to the difficulty of matching the bandwidth limits of the two methods. In addition, when shadowing or multiple scattering effects need to be accounted for, assessing surface texture parameters becomes much more complicated.

References

1. Lorentz, H.A.: Versuch einer Theorie der electrischen und optischen Erscheinungen in bewegten Körpern, E. J. Brill, Leiden (1895)
2. Drude, P.: Zur Elektronentheorie der Metalle, Part 1. Ann. Physik. **306**, 566–613 (1900)
3. Drude, P.: Zur Elektronentheorie der Metalle, Part 2. Ann. Physik. **308**, 369–402 (1900)
4. Brendel, R., Bormann, D.: An infrared dielectric function model for amorphous solids. J. Appl. Phys. **71**, 1–6 (1992)
5. Kim, C.C., Garland, J.W., Abad, H., Raccah, P.M.: Modeling the optical dielectric function of semiconductors: extension of the critical-point parabolic-band approximation. Phys. Rev. B. **45**, 11749–11767 (1992)
6. Tauc, J., Grigorovici, R., Vancu, A.: Optical properties and electronic structure of amorphous germanium. Phys. Status Solidi. **15**, 627–637 (1966)
7. Jellison Jr., G.E., Modine, F.A.: Parametrization of the optical functions of amorphous materials in the interband region. Appl. Phys. Lett. **69**, 371–373 (1996). Erratum, Appl. Phys. Lett. **69**, 2137 (1996)
8. Kronig, R.: On the theory of the dispersion of X-rays. J. Opt. Soc. Am. **12**, 547–557 (1926)
9. Kramers, H.A.: La diffusion de la lumière par les atomes. Atti Cong. Intern. Fisica (Transactions of Volta Centenary Congress, Como). **2**, 545–557 (1927)
10. Ferlauto, A.S., Ferreira, G.M., Pearce, J.M., Wronski, C.R., Collins, R.W., Deng, X., Ganguly, G.: Analytical model for the optical functions of amorphous semiconductors from the near-infrared to ultraviolet: applications in thin film photovoltaics. J. Appl. Phys. **92**, 2424–2436 (2002)
11. O'Leary, S.K., Johnson, S.R., Lim, P.K.: The relationship between the distribution of electronic states and the optical absorption spectrum of an amorphous semiconductor: an empirical analysis. J. Appl. Phys. **82**, 3334–3340 (1997)
12. Sellmeier, W.: Zur Erklärung der abnormen Farbenfolge im Spectrum einiger Substanzen. Annalen der Physik und Chemie. **219**, 272–282 (1871)
13. Cauchy, A.L.: Sur la réfraction et la réflexion de la lumière. Bulletin de Férussac, tomé. **14**, 6–10 (1830)
14. Cauchy, A.L.: Mémoire sur la Dispersion de la Lumière. J. G. Calve, Prague (1836)
15. Quinten, M.: A practical guide to optical metrology for thin films. Wiley-VCH, Weinheim (2012)
16. van de Hulst, H.C.: Light scattering by small particles. Dover Publications Inc., New York (1981)
17. Kerker, M.: The scattering of light. Academic, San Diego (1969)
18. Bohren, C.F., Huffman, D.R.: Absorption and scattering of light by small particles. Wiley, New York (1983)
19. Mishchenko, M.I., Hovenier, J.W., Travis, L.D. (eds.): Light scattering by nonspherical particles. Academic, San Diego (2000)
20. Mishchenko, M.I., Travis, L.D., Lacis, A.A.: Scattering, absorption, and emission of light by small particles. Cambridge University Press, Cambridge (2002)
21. Quinten, M.: Optical properties of nanoparticle systems – Mie and beyond. Wiley-VCH, Berlin (2011)
22. Beckmann, P., Spizzichino, A.: The scattering of electromagnetic waves from rough surfaces. Pergamon Press, New York (1963)

23. Stover, J.C.: Optical scattering – measurement and analysis, 2nd edn. SPIE Optical Engineering Press, Bellingham (1995)
24. Ogilvy, J.A.: Theory of wave scattering from random rough surfaces. Hilger, Bristol (1991)
25. Nicodemus, F.E., Richmond, J.C., Hsia, J.J., Gingsberg, I., Limberis, T.: Geometric Considerations and Nomenclature of Reflectance NBS Monograph 160. U. S. Department of Commerce, Washington, DC (1977)
26. Rayleigh, L.: The Theory of Sound, 3rd edn. Macmillan, London (1896)
27. Fano, U.: The theory of anomalous diffraction gratings and of quasi-stationary waves on metallic surfaces (Sommerfeld's waves). J. Opt. Soc. Am. **31**, 213–222 (1941)
28. Rice, S.O.: Reflection of electromagnetic waves from slightly rough surfaces. Commun. Pure Appl. Math. **4**, 351–378 (1951)
29. Bruce, N.C., Dainty, J.C.: Multiple scattering from random rough surfaces using the Kirchhoff approximation. J. Mod. Opt. **38**(3), 579–590 (1991)
30. Bruce, N.C., Dainty, J.C.: Multiple scattering from rough dielectric and metal surfaces using the Kirchhoff approximation. J. Mod. Opt. **38**(8), 1471–1481 (1991)
31. Bruce, N.C.: Kirchhoff calculations of the coherent scatter from a series of very rough surfaces. Appl. Opt. **34**(24), 5531–5536 (1995)
32. Harvey, J.E.: Light-Scattering Characteristics of Optical Surface, PhD thesis, University of Arizona, 1976; adviser: R. V. Shack
33. Krywonos, A.: Predicting Surface Scatter using a Linear Systems Formulation of Non-Paraxial Scalar Diffraction, PhD thesis, University of Central Florida, 2006. Adviser: J. E. Harvey
34. Davies, H.: The reflection of electromagnetic waves from a rough surface. Proc. Inst. Elec. Eng. **101**, 209–214 (1954)
35. Church, E.L., Jenkinson, H.A., Zavada, J.M.: Relationship between surface scattering and microtopographic features. Opt. Eng. **18**(2), 125–136 (1979)
36. Bennett, H.E., Porteus, J.O.: Relation between surface roughness and specular reflection at Normal incidence. J. Opt. Soc. Am. **51**, 123–129 (1961)
37. Bennett, J.M., Mattsson, L.: Introduction to Surface Roughness and Scattering, 2nd edn. Optical Society of America, Washington, DC (1999)
38. Harvey, J.E., Schröder, S., Choi, N., Duparré, A.: Total integrated scatter from surfaces with arbitrary roughness, correlation widths, and incident angles. Opt. Eng. **51**(013402), 1–11 (2012)
39. ISO 10110-8: Optics and photonics – preparation of drawings for optical elements and systems – part 8: surface texture. International Organization for Standardization (2010)
40. SEMI MF 1048-1109: Test method for measuring the effective surface roughness of optical components by total integrated scattering. Semiconductor Equipment and Materials International (2009)
41. SEMI ME 1392-1109: Guide for angle resolved optical scatter measurements on specular or diffuse surfaces. Semiconductor Equipment and Materials International (2009)

Chapter 5
Optical Surface Metrology: Methods

Abstract The techniques in surface metrology with the largest diversity are based on electrodynamics, more precise on the interaction of electromagnetic waves with the surface of the workpiece. State-of-the-art camera technology, proper attention to the internal optical design, light sources, and computing capabilities have led to impressive optical metrology systems with considerable specifications. The most relevant techniques are discussed in the following Sects. 5.1, 5.2, 5.3, 5.4, 5.5, 5.6, 5.7, 5.8, 5.9, and 5.10.

This chapter comprises the following methods:

- Confocal Optical Profiling,
- Light Sectional Methods,
- Various Microscopy Methods,
- Various Interferometric Methods,
- Wave Front Sensing,
- Deflectometry,
- Elastic Light Scattering, and
- Spectral Analysis and Characterization.

5.1 Chromatic Confocal Surface Profiling

A simple but effective measuring technique has been established in optical profilometry for surface topography and roughness measurement: the measurement with a *chromatic white light sensor* [1–4]. This sensor utilizes the chromatic aberration (see again Sect. 4.8) of the lens system of the measuring head to measure the distance sensor – surface of the object without moving the measuring head in *z*-direction. In principle, it is a reflectometer with aberration optics.

The *chromatic aberration* is usually mostly unfavorable in optical applications. Rather, in the chromatic confocal sensor it is utilized to establish a robust wearless sensor. Its principle is sketched in Fig. 5.1.

M. Quinten, *A Practical Guide to Surface Metrology*, Springer Series in Measurement Science and Technology,
https://doi.org/10.1007/978-3-030-29454-0_5

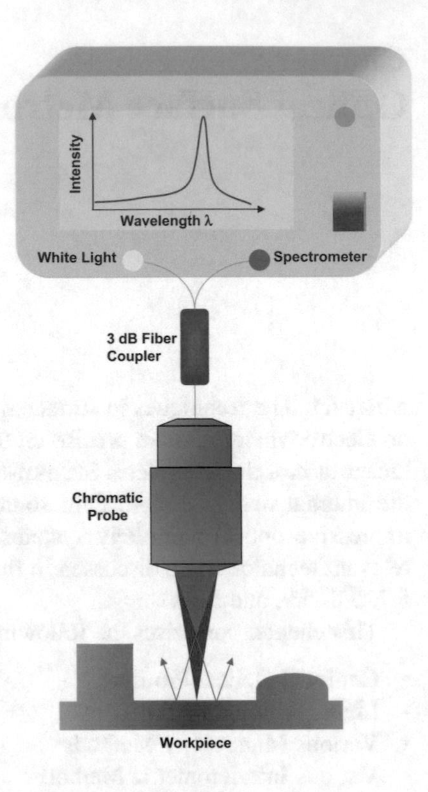

Fig. 5.1 Principle of a chromatic white light sensor

The light of a white light source (typically a white light LED) is coupled into an optical fiber (mostly with core diameter $d = 50$ μm). The end of the fiber is focused with the chromatic probe in a spot on the specimen. The chromatic probe has a magnification $M < 1$, mostly $M = 0.1$. The chromatic probe consists of a set of achromatic lenses for imaging and correction of spherical aberrations and a chromatic lens as front lens that establishes the measuring range. The chromatic aberration is realized either with a spherical or aspherical lens with high refractive index and high dispersion. The measuring range is defined by the difference in the focal lengths for the shortest and the longest wavelength in the useful spectral range. Typical values for the measuring range of chromatic white light sensors used in surface topography measurement are between 100 and 3000 μm. The mostly used chromatic white light sensors have an aperture angle of $\alpha = 30°$ or $\alpha = 45°$.

Measuring the light reflected at the specimen in the focal point of the front lens with a miniaturized spectrometer one obtains a reflectance spectrum that is peaked at the wavelength which is in focus on the sample. The other wavelengths get spatially

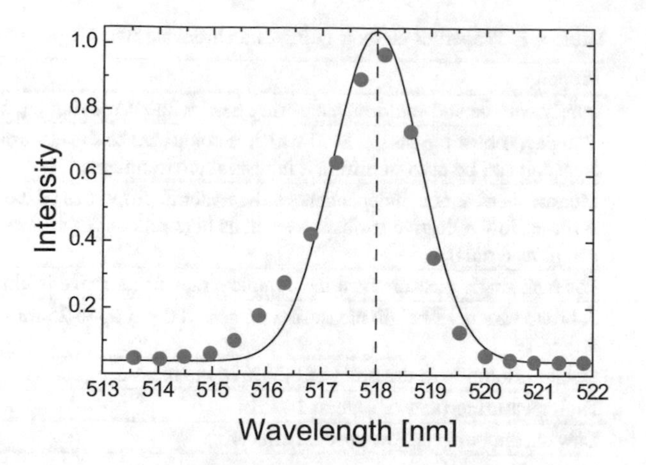

Fig. 5.2 Intensity maximum in the reflectivity measured with a chromatic white light sensor. The solid line represents the fit on the data for finding the peak wavelength position

distributed since they are not in focus resulting in lower intensities in the spectrum. This is illustrated in Fig. 5.2.

The determination of the wavelength where the measured reflectance is peaked is done with a fit (mostly parabola or Gaussian) on the intensity curve or in calculating the center of gravity of the peak area, e.g. the portion above 50% of the peak intensity. Both methods result in sub-pixel accuracy on the peak abscissa. The latter requires less computational resources and can easier be implemented in the embedded electronics of commercial probes. The calibration of the peak wavelength position to the distance allows to determine distances and step heights with nanometer resolution, depending on the measuring range of the used chromatic white light sensor.

For a correct determination of the peak wavelength it is necessary to calibrate the spectrometer. First of all, the dark current signal must be determined. The dark current signal originates from temperature released electrons in the detector and forms a constant signal level that cannot be used for evaluation but must be subtracted from each measurement. In addition, the spectrum of the light source must be determined once as reference spectrum. As it is also contained in each measured spectrum, the measured spectrum can be corrected by division through the reference spectrum.

The spot size of the chromatic confocal probe has a direct influence on the lateral resolution of the measurement. The achievable spot size depends upon the numerical aperture of the front lens, the mean focal distance of the front lens, the magnification of the chromatic probe, and the core diameter of the optical fiber. Typical spot sizes of commercial probes are $\leq$5 µm for measuring ranges $\leq$1 mm and between 5 and 15 µm for larger vertical ranges between 2 and 25 mm. As the intensity spread function inside a spot can be described by a Bessel function the lateral resolution is approximately half of the spot size.

Table 5.1 Properties an their rating of a chromatic white light sensor

Property	Rating
Purely passive and wearless measuring head without moving parts	+
Compact robust measuring head which is connected to the electronics only by a fiber. Hence, it can be used in different industrial environments	+
Measurement almost independent of the material properties of the specimen (colored surfaces, low reflective surfaces as well as high reflective surfaces, opaque and transparent materials)	+
For each single measurement the complete measuring range is utilized	+
A broad variety of height measuring ranges: 100 µm up to 25 mm, most common 600–3 mm	+
Height resolution in the order of 0.33% of the measuring range	+
High measuring rates of several 10 kHz	+
Low shading effects, low speckle effects	+
Easy to integrate	+
Lateral measuring range amounting up to several 100 mm, as it is possible to setup the length of a profile	+
Lateral resolution restricted by achievable focus spot size	–
Sensitive at edges with resulting artifacts	–
Transparent materials of thickness less than approximately 1/10 of the measuring range are critical due to the reflection at the rear side	–

If the sample is curved the maximum local slope can have maximum an angle in the order of the aperture angle of the chromatic probe. For larger angles the reflected light will not be detected. For a rough surface of the manufactured workpiece light will still enter the aperture of the probe also at larger slope angles due to diffuse scattering. In dependence upon the amount of this scattered light a distance can be evaluated from the reflectance spectrum even for larger slope angles.

As the sensor measures only at one point either the sensor or the specimen must be moved to obtain a 2D line profile analogous to the tactile measurement or even a 3D measurement by scanning a x-y plane. When moving to another point of the surface another wavelength may be in focus resulting in a peak in the spectrometer at this new wavelength. This allows for determination of the topography of the surface within the lateral resolution of the used chromatic probe.

This chromatic confocal sensor has some advantages and disadvantages. All "plus" and "minus" are summarized in Table 5.1.

The chromatic white light sensor is a robust wearless sensor and evolved to the workhorse in optical surface topography measurement analogous to the stylus tip. The broad variety of measuring ranges allows for a wide range of applications. Therefore, it is used widespread from many companies. An extension of the working principle of the chromatic confocal profilometer was already done in 2006 with the combination with an interferometer [5] but without further development or realization in a commercial instrument.

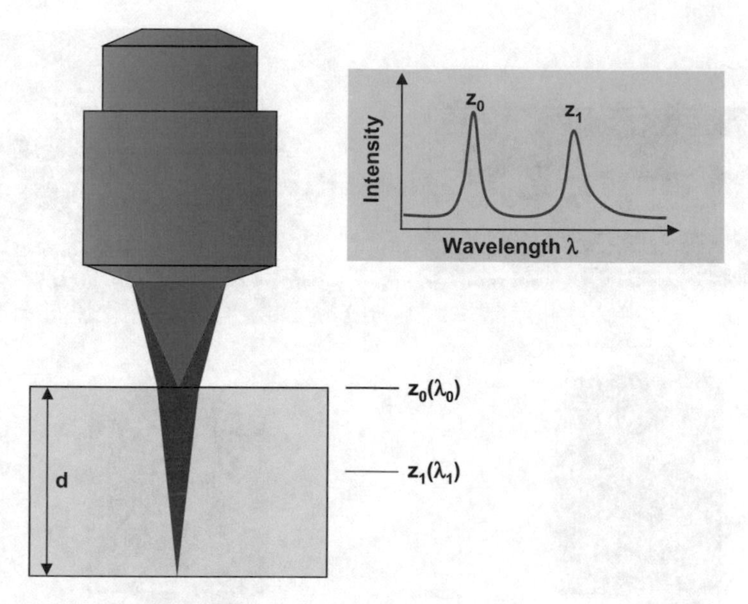

Fig. 5.3 Measuring the thickness of transparent workpieces or transparent coatings with the chromatic white light sensor

For transparent workpieces, e.g. glass sheets or with a transparent coating, the thickness of the workpiece or of the coating can be determined from distance measurement with the sensor. The principle is quite simple. For both interfaces the wavelength which is focused on the corresponding interface yields a peak in the spectrometer. This is illustrated in Fig. 5.3. It is however fallacious to believe that the measured distance Δz between these two peaks directly corresponds to the thickness. Due to the refraction of the incident light the foci of the penetrating waves are prolonged. This must be considered when calculating the thickness from the peak positions. A simple calculation using geometric relations and Snell's law leads to a relation between the actual thickness d of the transparent workpiece or the transparent film and the distance $\Delta z = z_1 - z_0$ obtained from the evaluation of the distance signal of the chromatic white light sensor

$$\mathrm{d} = \Delta \mathrm{z} \cdot \mathrm{n} \cdot \frac{\cos\left(\sin^{-1}\left(\frac{\sin\alpha}{\mathrm{n}}\right)\right)}{\cos\alpha} \tag{5.1}$$

where n is the refractive index of the transparent material. In a first approach the ratio of the cosine functions is approximately 1 and the thickness d is simply obtained by multiplication of the measured Δz with the refractive index n. For a more correct result for d one has to integrate this equation over the full aperture as discussed in [6].

In Fig. 5.4 first a set of pictures of chromatic confocal distance sensors is shown. The pictures are courtesy of Precitec-Optronik GmbH, Neu-Isenburg, Germany.

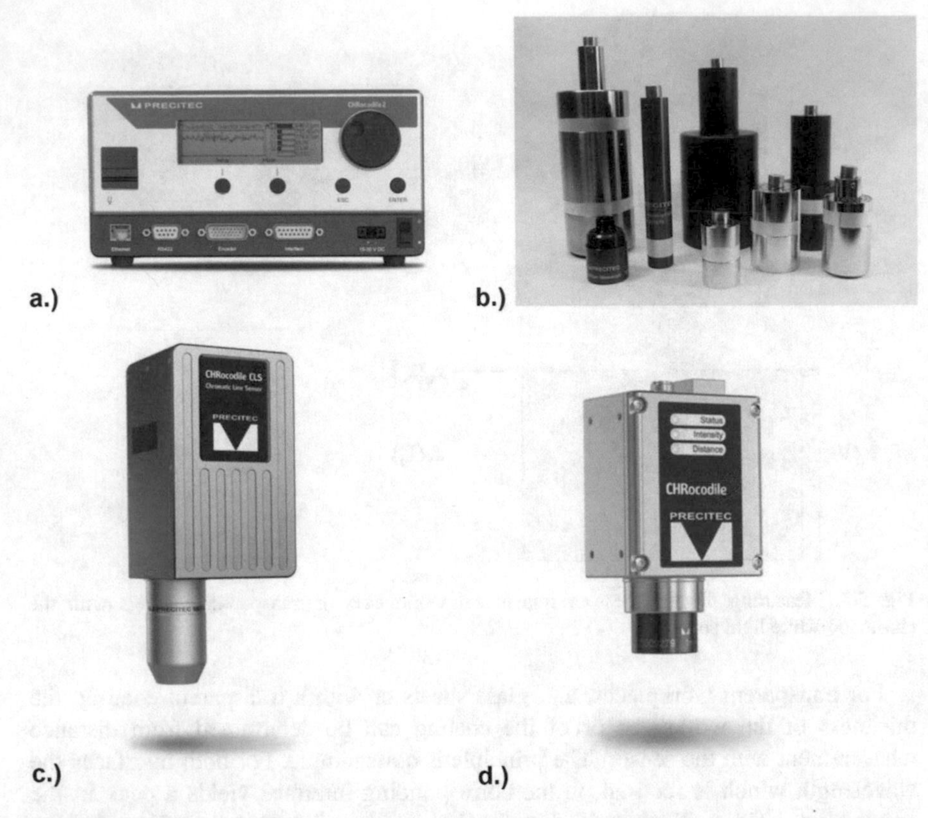

Fig. 5.4 (**a**) Electronics (light source, spectrometer, display) of a chromatic confocal sensor, (**b**) Chromatic probes with various measuring ranges, (**c**) Multiple point chromatic white light sensor (192 focus points in a line, electronics and optics all inclusive), (**d**) Compact chromatic white light sensor with electronics and optics all inclusive. (Courtesy of Precitec Optronik, Neu-Isenburg, Germany)

Then, some examples of measured surface topographies follow in Figs. 5.5 and 5.6. These pictures are courtesy of FRT GmbH, Bergisch Gladbach, Germany. In Fig. 5.5 the measurements were taken with a single chromatic white light sensor with the measuring range being different from picture to picture. In Fig. 5.6 the measurements were conducted with a multiple point chromatic white light sensor.

The examples in Fig. 5.5 clearly demonstrate the multiplicity of the chromatic white light sensor. The broad variety of measuring ranges allows for high resolution areal measurements on micro systems as micro prisms in Fig. 5.5a or MEMS microfluidics components in Fig. 5.5c as well as measurement of extended areas of several millimeters as shown in Fig. 5.5b, d with topography measurements of a test wafer with various test structures or the surface of a Fresnel lens.

For high speed areal topography measurements, a multiple point chromatic white light sensor is an economic alternative if the lateral resolution can be reduced. This is

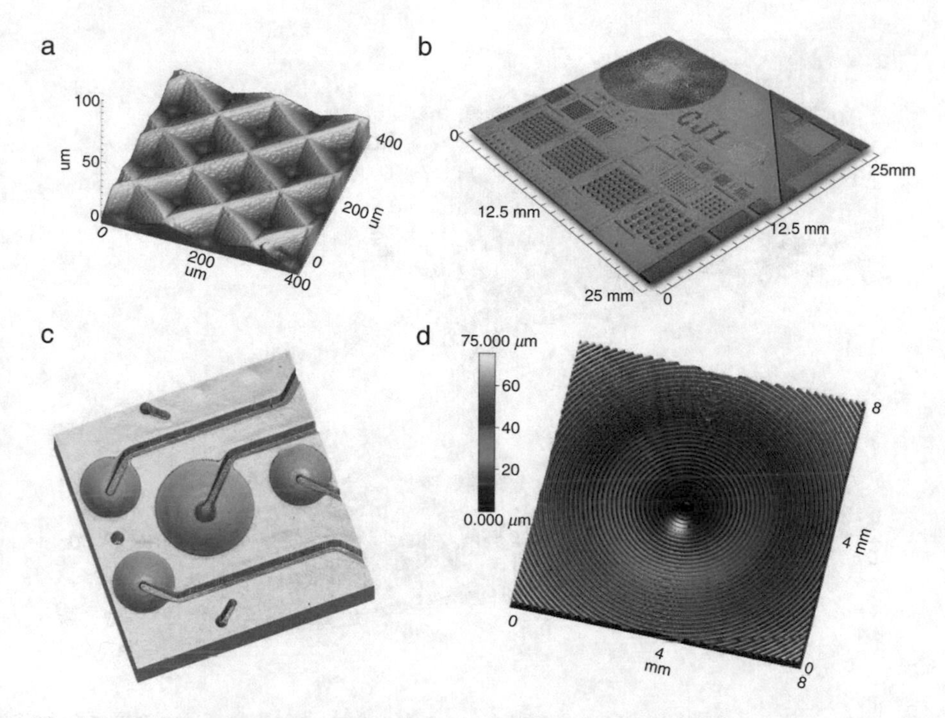

Fig. 5.5 Areal measurements with the chromatic white light sensor FRT CWL: (**a**) topography of micro prisms, (**b**) topography of a test wafer with various test structures, (**c**) topography of a MEMS microfluidics component (**d**) topography of a Fresnel lens. (Courtesy of FRT GmbH, Bergisch Gladbach, Germany)

demonstrated with the examples in Fig. 5.6 with measurements on artificial leather and on a wafer with solder balls. The measured areas have several millimeters in length and width but the measuring time is pretty small with 0.2 s and 0.4 s due to the simultaneous measurement with 192 focus points along a line.

Another often used application of the chromatic white light sensor is the measurement with two opposite sensors for simultaneous measurement on the top side and the bottom side of the workpiece. This twin system must be calibrated with a certified end-gauge block. For nontransparent workpieces like semiconductor wafers this enables the measurement of the thickness of the wafer, its total thickness variation (TTV), the bow and warp of the wafer, and many other relevant quantities according to SEMI standards. For transparent workpieces like optical elements both surfaces can be inspected simultaneously. This is of particular interest if both sides are different, e.g. the top side being an asphere and the bottom side being a segmented free form optics. An example is shown in Fig. 5.7. The pictures are courtesy of FRT GmbH, Bergisch Gladbach, Germany.

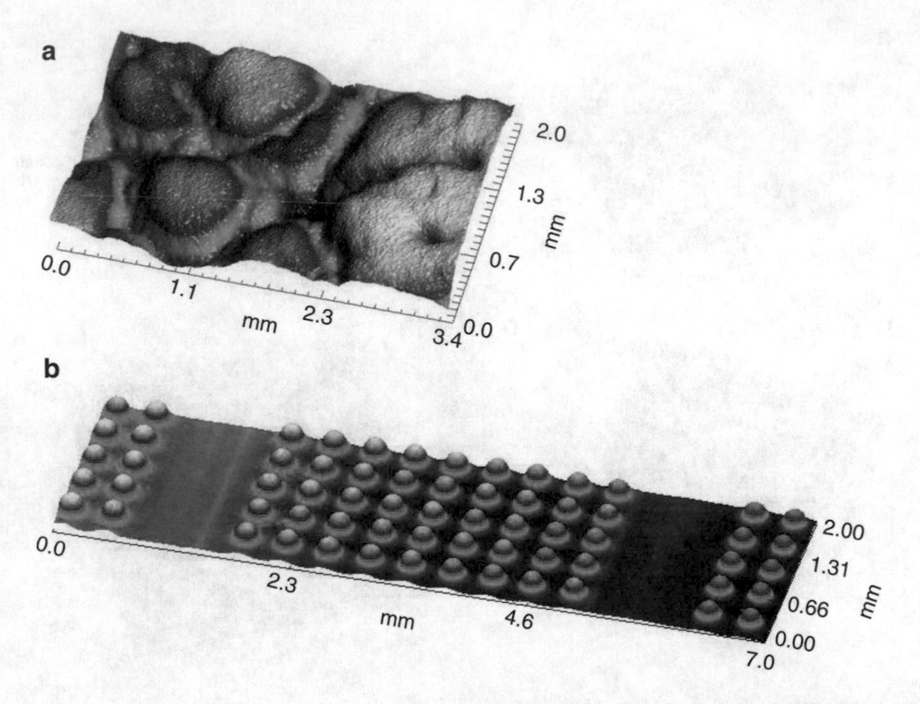

Fig. 5.6 Areal measurements with the multiple point chromatic white light sensor FRT SLS: (**a**) artificial leather, measurement time 0.2 s, (**b**) wafer with solder balls, measurement time 0.4 s. (Courtesy of FRT GmbH, Bergisch Gladbach, Germany)

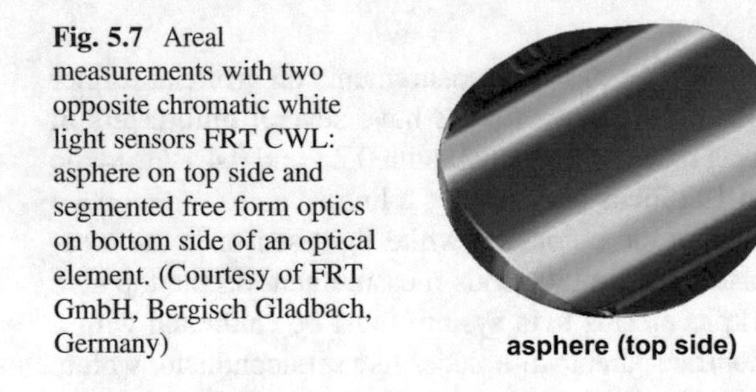

Fig. 5.7 Areal measurements with two opposite chromatic white light sensors FRT CWL: asphere on top side and segmented free form optics on bottom side of an optical element. (Courtesy of FRT GmbH, Bergisch Gladbach, Germany)

5.2 Surface Profiling with an Autofocus Sensor

The laser *autofocus sensor* has attained a wide distribution in the compact disc technique, for DVDs, and for magneto-optical data storage units as reading unit since about 40 years. Its first mention was in the Philips technical review from 1973 [7]. Since this time the number of principles of focus point detection increased (see for example [8]). The autofocus sensor has become also very popular for the geometrical measurement of various specimens. With an autofocus sensor a laser beam gets automatically focused on the surface and the reflected light is detected

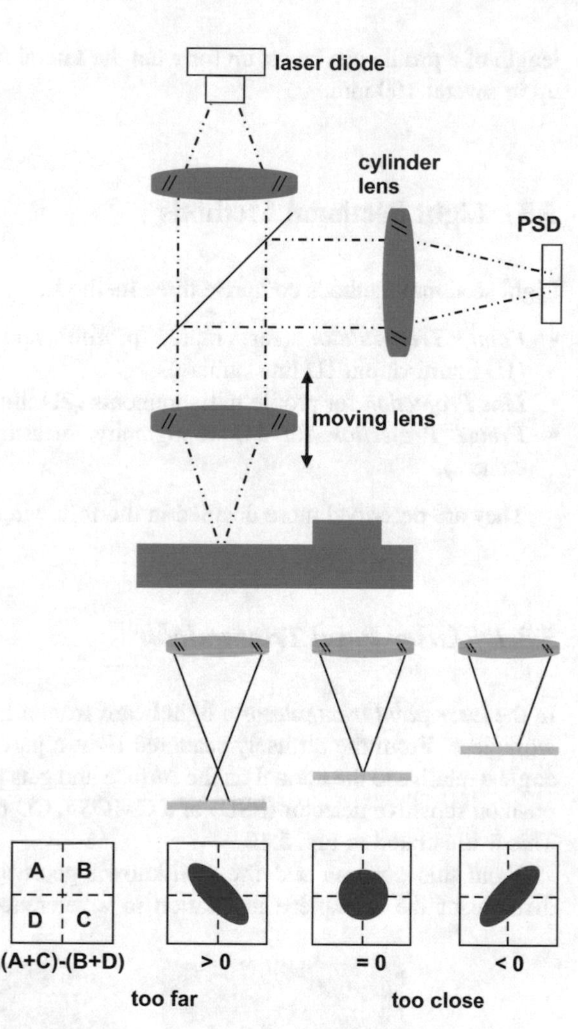

Fig. 5.8 Sketch of the principle of an autofocus sensor

Fig. 5.9 Detection spot on the PSD, left: out of focus, far distance; mid: in focus, right: out of focus, close distance

with a position sensitive detector (four-quadrant diode) using the so-called astigmatism detection [9, 10]. The working principle is sketched in Figs. 5.8 and 5.9.

Dividing the four-quadrant diode in A, B, C, and D one can distinguish with (A+C) – (B+D) three cases. If the surface is in focus in a point (x, y) the detection spot is circular and (A+C) – (B+D) = 0. If the surface is out of focus in this point the detection spot is deformed by the cylinder lens to an ellipse with its long axis along the diagonal from right to left or from left to right. Then, (A+C) – (B+D) > 0 or (A+C) – (B+D) < 0 indicating that the sensor is too far or too close.

The focusing lens gets updated in a closed loop so that the focus remains on the surface. From the amount of tracking the height z at position (x, y) is retrieved.

This technique is a smart method for surface metrology with a high height resolution of 20 nm. Problems arise with steep edges and with holes. Overshots and undershots may occur because the surface must be permantly searched for. The length of a profile can be set up for what the lateral measuring range can amount to up to several 100 mm.

5.3 Light Sectional Methods

Light sectional methods comprise three methods:

- *Point Triangulator* for rapid profile and topography measurements (1D illumination, 1D line camera),
- *Line Projection* for profile measurements (2D illumination, 2D camera),
- *Fringe Projection* for 3D topography measurement (3D illumination, 2D camera).

They are described more detailed in the following three Sections.

5.3.1 Laser Point Triangulator

In the *laser point triangulator* a light beam from a LED or a laser is directed to the workpiece. From the diffusely scattered light a part gets collected under a distinct angle α relative to the normal on the surface and gets projected with an objective on a position sensitive detector (PSD) or a CMOS/CCD detector, mostly a line detector. This is illustrated in Fig. 5.10.

From this position and the well-known geometry of the imaging system the distance of the workpiece in relation to a reference distance can be determined.

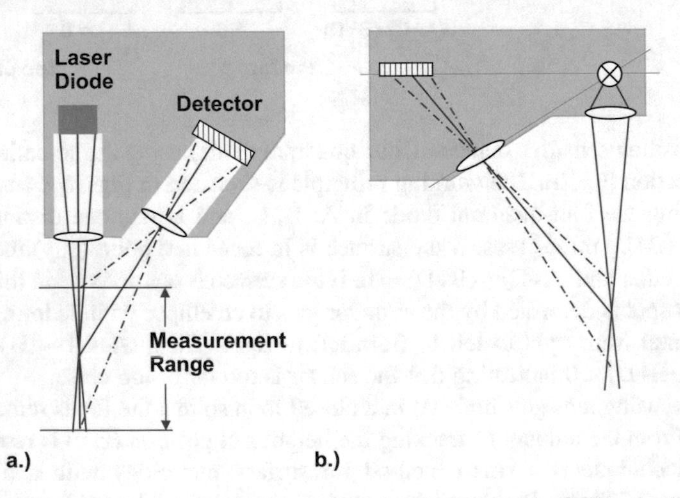

Fig. 5.10 (**a**) Principle of a point triangulator, (**b**) Point triangulator that fulfills the Scheimpflug condition

When moving now the workpiece along the beam axis by Δz the position of the image point on the PSD also changes by Δx

$$\Delta x = \beta \cdot \sin(\alpha) \cdot \Delta z \tag{5.2}$$

where β is the reproduction scale of the used optics. The proportionality factor $\beta \cdot \sin(\alpha)$ is approximately constant. Deviations can be included by calibration.

Point triangulators have an ideal operating point sometimes referred to as the *standoff distance*. At this point the incident light is at its sharpest focal point and the reflected light is in the center of the PSD. When moving the target, this spot moves towards the ends of the PSD defining the measuring range of the detector. Under certain circumstances it may be necessary to have the light spot on the surface being imaged always sharp independently of the distance of the workpiece. To achieve this the triangulator is built so that the beam axis, the objective plane, and the image plane intersect in a point as indicated by the green lines in Fig. 5.10b. This is called *Scheimpflug condition* according to Theodor Scheimpflug (1865–1911) [11].

High frequency modulation of the light source allows to suppress interfering light as well as to sample rapidly moving objects (measuring rates of several kHz). The measuring range of a point triangulator can amount from a few millimeters to a few meters depending on the application. The accuracy in vertical direction amounts to a few percent of the measuring range. It is a function of linearity, resolution, temperature stability, and drift, with linearity being the major contributor. In lateral direction the accuracy is mainly determined by the spot size on the specimen. Color and brightness of the specimen have an important impact on the magnitude of the scattered light. Very smooth surfaces with almost no scattering cannot be measured with triangulation. Strong fluctuations within the measuring spot cause systematic errors. For workpieces exhibiting textures it is favorable to position the sensor relative to the texture so that the texture is perpendicular to the plane of the triangulation. Inclination of the specimen may lead to errors. They can be minimized by positioning the workpiece parallel to the plane of triangulation. Shadowing effects are challenging for point triangulators. They can be reduced by a small triangulation angle. The primary factor in determining resolution is the electrical noise of the system.

When positioning two triangulator sensors opposing they can be used to determine the thickness of nontransparent materials by measuring the distance of the top sensor to the top surface and the distance of the bottom sensor to the bottom surface. This twin system must be calibrated with a certified end-gauge block.

5.3.2 Line Projection

The *line projection* is an extension of the point triangulator on a line on the workpiece (see Fig. 5.11). The light beam gets expanded with a special optics to a line. This enables the fast measurement of profiles with a field of view of up to about

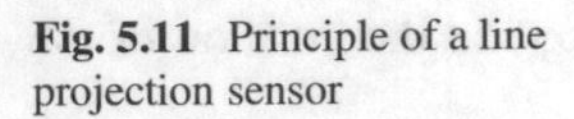
Fig. 5.11 Principle of a line projection sensor

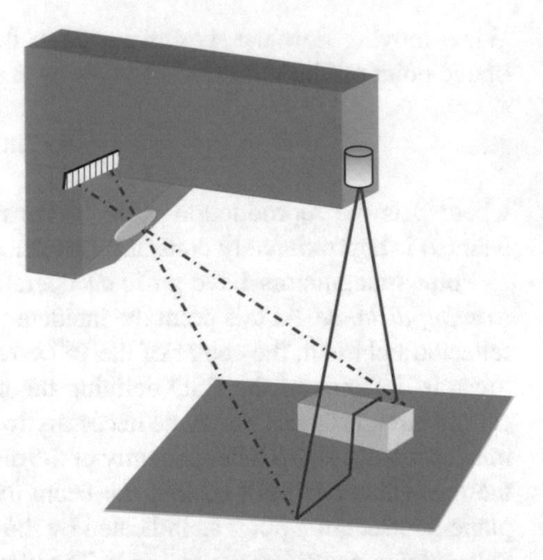

1300 mm. For the detection of the diffusely scattered light the line detector is replaced by an array detector (camera). The set-up of the sensor fixes the working distance, the measuring range, and the resolution. A measurement range up to 800 mm is possible. The maximum vertical resolution amounts to 1/5000 of the measuring range. The sensor fulfills the Scheimpflug condition.

As mostly lasers or laser diodes are used for illumination, disturbing speckles are unavoidable. They can be reduced by using laser light sources with reduced coherence length or with large apertures of the camera objective as far as this is possible with respect to the required depth of field.

5.3.3 Fringe Projection

Fringe projection is an optical technique that enables fast and area-wide recording of the surface topography. A modern fringe projection system consists of two digital cameras and a fringe projector in a fixed setup. Cameras and fringe projector enclose a fixed angle. Figure 5.12 shows a typical setup of a fringe projection system and a picture of the commercially available fringe projection system ATOS Triple Scan from GOM GmbH, Braunschweig, Germany.

A sequence of shifted patterns of parallel sinusoidal stripes often combined with gray code sequences gets projected on the surface of the object. If the surface is not planc the stripes pattern gets distorted. Both cameras record an image of the distorted fringe pattern. From the analysis of the recorded pattern corresponding object points can be identified in both cameras. The corresponding points are pixels in both cameras that look at the same object point. These pixels and the object point form a triangle with angles α and β and the base length *b* from which the 3D coordinates of the object point are calculated. The identification of the corresponding object points

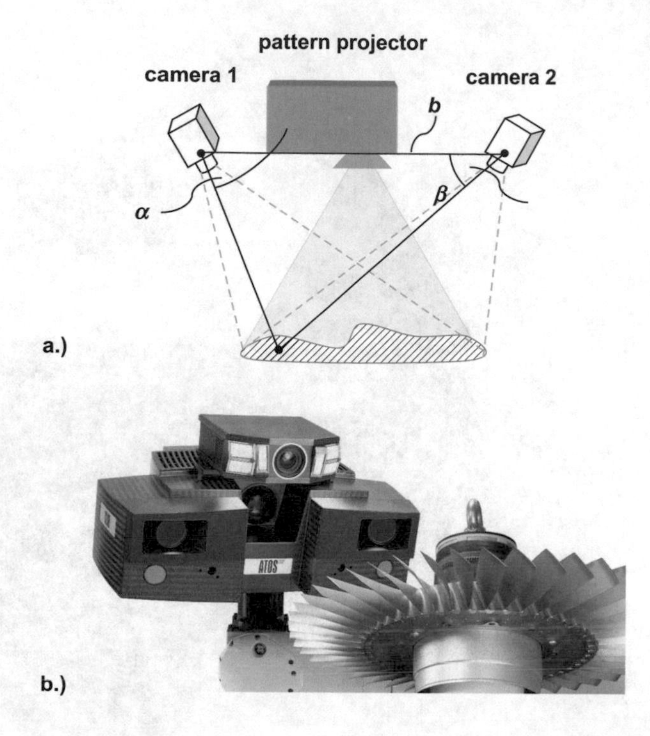

Fig. 5.12 (**a**) Sketch of a modern fringe projection system with two cameras, (**b**) Commercially available system ATOS Triple Scan of GOM GmbH, Braunschweig, Germany. (Picture courtesy of GOM GmbH)

can be made with a phase correlation method [12] or a temporal correlation of image sequences of statistical patterns for high speed measurements [13]. For the generation of the statistical patterns various possibilities were developed [14–16]. The use of gray code sequences and phase-coded methods requires accidential structures or reference marks on the object to assign an object point unambiguously to the corresponding pixels on the two cameras. Thus, they are well-suited for stationary objects. For moving objects as in inline measurements frequency-coded methods are more favorable which on the other hand are less robust against surface texture effects.

The lateral resolution is given by the field of view divided by the number of pixels in x- and y-direction. The triangulation angles α and β and the accuracy in determination the phase of the reflected pattern determines the resolution in vertical direction that amounts up to 1/10,000 of the vertical measuring range. The use of two cameras remarkably improves the quality of the data with respect to former fringe projection systems with only one camera. The shift of the pattern allows to calculate planes that lie narrower than the fringes in the single projected pattern. Beam paths of both cameras and the projector are calibrated in advance.

The following Fig. 5.13 shows two measurements with the ATOS Triple Scan of GOM GmbH in the automotive industry. The first picture depicts the inspection of a trunk lid. The numbers are deviations from the set values. The second picture shows the inspection results of parts of a car body.

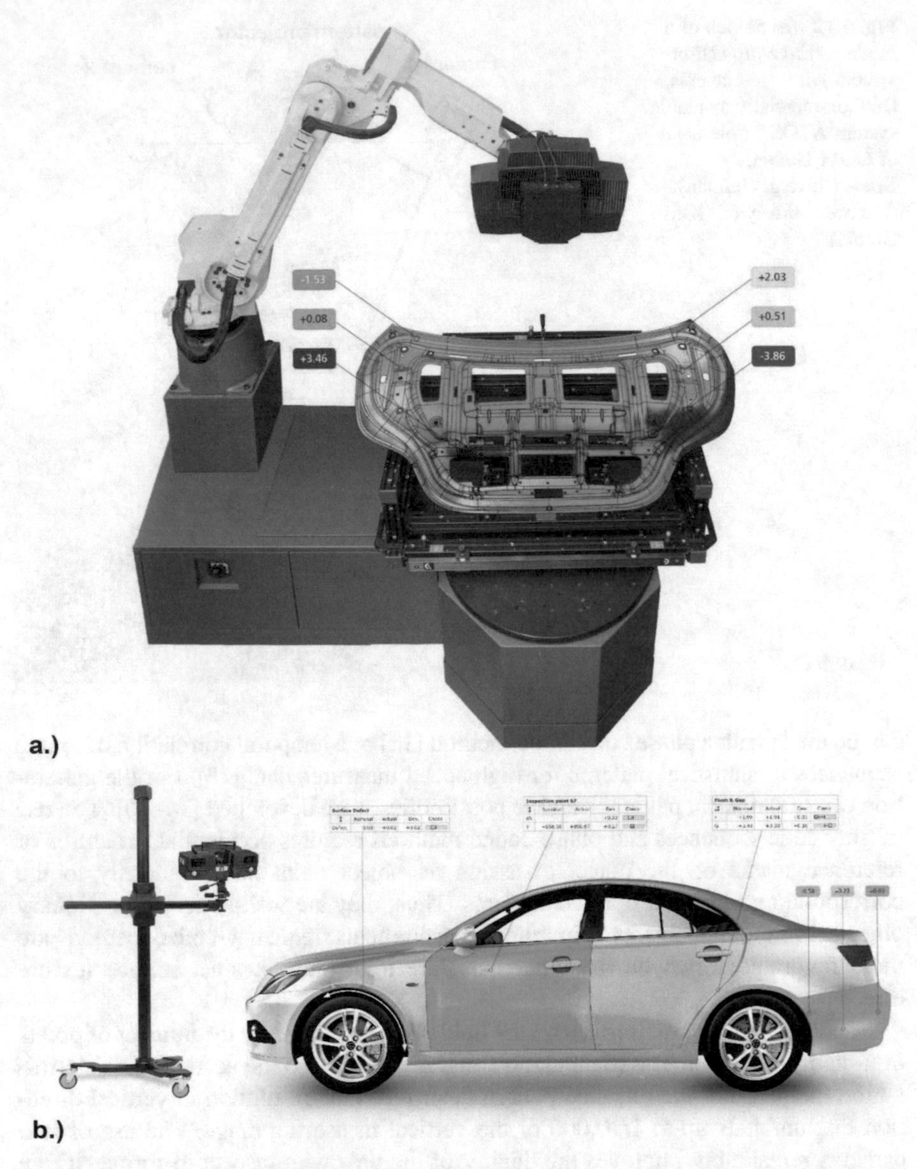

Fig. 5.13 Measurement results with the ATOS Triple Scan: (**a**) inspection of a trunk lid, (**b**) inspection of parts of a car body. (Pictures are courtesy of GOM GmbH, Braunschweig, Germany)

5.4 Microscopy Methods

The microscope is an optical instrument that allows an enlarged view (greek: skopein – to view sth.) on a small (greek: mikrós – small, minute) object. This is enabled by a two-stage optical imaging using a microscope objective and a microscope ocular or eyepiece. Pioneers of modern microscopy were Robert Hooke (1635–1703) and Antoni van Leeuwenhoek (1632–1723). In 1665 Hooke constructed for the first time a microscope that consisted of many lenses. With his microscope he could see cells of suberic mesh. Leeuwenhoek was the only at that time who could manufacture lenses as exact as he arrived in a magnification of $M = 270$. In 1685 he even could observe bacteria. Also modern microscope objectives consist of many lenses. This kind of construction allows for a high magnification and improved image quality.

But the story of microscopy does not end here. Rather, several developments have been carried out to improve contrast, vertical resolution, and lateral resolution by use of special optical elements, special techniques, or highly sophisticated evaluation algorithms. In this Sect. the main relevant techniques for surface metrology basing upon the classical microscope are presented besides the microscope itself.

5.4.1 Classical Microscopy

In a classical microscope the objective generates a magnified real intermediate image in the tube. The eyepiece usually acts as magnifying glass that enables a view on the intermediate image with a larger visual angle. This is achieved by generating a virtual image in front of the ocular that is usually observed with the human eye. The eye lens generates a real image from this virtual image on the retina. In many modern applications a camera is used for observation instead of or in addition to the human eye. Then, either the eyepiece must be shifted so that the intermediate image lies in between the focal length f_{oc} of the ocular and $2f_{oc}$ to generate a magnified real image behind the eyepiece on the camera detector area. Or the camera is equipped with a focusing lens and acts similar to the human eye with eye lens and retina. The optical path in a classical microscope with camera as image detector is shown in Fig. 5.14. The tube lens is not necessary but helpful for keeping the tube dimensions small without disturbing the imaging.

Since its invention the classical optical microscope and its successive variants have generally relied on the objective lens as key element. But the quality of

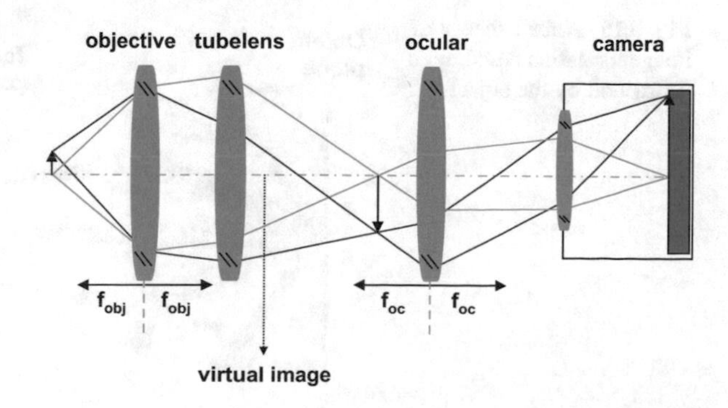

Fig. 5.14 Optical path in the classical microscope with camera as image detector

objectives was rather modest up to the middle of the nineteenth century and the production was time-consuming and expensive. Improvements were made by William Hyde Wollaston, Sir David Brewster, Andrew Pritchard, Joseph von Fraunhofer, and many others. The breakthrough succeeded Ernst Abbe who was invited by Carl Zeiss to develop new microscope objectives. On the base of physical optics he developed first the theory of image formation in 1873 [17] and then calculated new objectives. Finally, he also established a modern production of objectives by the construction of measurement devices (thickness gauge, refractometer, spectrometer) for the economical manufacturing of high-quality objectives. In cooperation with Friedrich Otto Schott he also developed new glasses with well-defined optical material constants which clearly improved the microscopes of Carl Zeiss. Today the objective lens is a multi-element lens, thus the number of lenses in a modern microscope can easily exceed 20.

The lateral resolution of a microscope is theoretically limited by the wavelength λ (usually the mid wavelength of the used spectral range, for the visible spectral region $\lambda = 560$ nm) and the *numerical aperture NA* of the objective

$$\mathrm{A} = \frac{1.22 \cdot \lambda}{2 \cdot \mathrm{NA}} \tag{5.3}$$

This condition follows from Abbe's theory of image formation. In this theory the object is assumed to diffract the light similar to a diffraction grating. For reconstruction of the image of the object at least the first diffraction order of each object point is necessary beyond the zeroth order (see Fig. 5.15). Then, the maximum ascertainable angle is given by the angle of first order diffraction whereby it follows for the numerical aperture

$$\mathrm{NA} = \mathrm{n} \cdot \sin(\alpha) \tag{5.4}$$

with n being the refractive index of the medium between object and the microscope objective (mostly air with $n = 1$).

Each point of the object has an image in the image plane. Due to diffraction the image is however not a point but has an intensity distribution that can be described

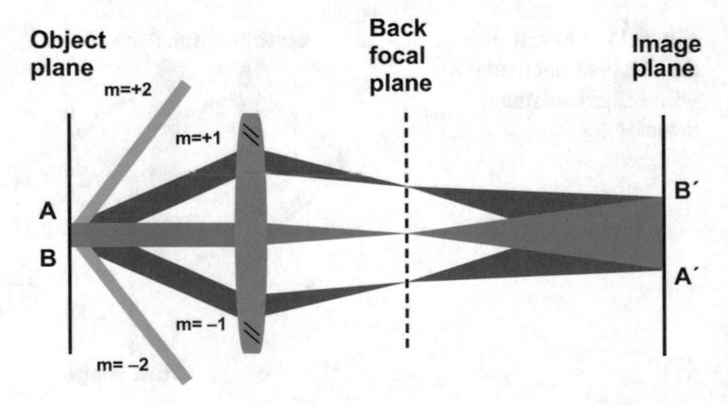

Fig. 5.15 Abbe's theory of image formation based upon diffraction by the object

by the function $(J_1(z)/z)^2$ where $J_1(z)$ is the Bessel function of first order. This distribution has zeros and maxima. According to Rayleigh's criterion two points can clearly be distinguished when the first maximum of the intensity distribution of point 1 lies in the first zero of the intensity distribution of point 2. From this condition Eq. (5.3) follows directly as the first zero of $J_1(z)$ is at $z = 3.8317$.

The vertical resolution of a microscope or more general of each imaging lens system is also limited, namely by the *depth of field* DoF

$$\text{DoF} = \frac{n \cdot \lambda}{2 \cdot (\text{NA})^2} + \frac{340\,\mu\text{m}}{\text{NA} \cdot M_{\text{TOT, VIS}}} \tag{5.5}$$

with $M_{TOT,VIS}$ being the total magnification of the microscope in the visible spectral range. Here, the length of 340 μm results from the resolving power of the eye. This formula is known as *Berek's formula* [18]. When using a camera or another detector, the equation reads

$$\text{DoF} = \frac{n \cdot \lambda}{2 \cdot (\text{NA})^2} + \frac{n \cdot e}{\text{NA} \cdot M} \tag{5.6}$$

with the variable *e* being the smallest distance that can be resolved by a detector that is placed in the image plane of the microscope objective whose lateral magnification is *M*. The first term in both equations is known as *Rayleigh's depth of field*. The depth of field of a microscope is typically in the order of less than 1–2 μm.

What is the general meaning of the depth of field for imaging? To give an answer to this question the imaging with a single lens with fixed positions of the lens and the image plane is considered in Fig. 5.16.

The first case considers that the object point is at a distance s_o in front of the lens so that the image point lies exactly in the image plane at a distance s_i. Next, the

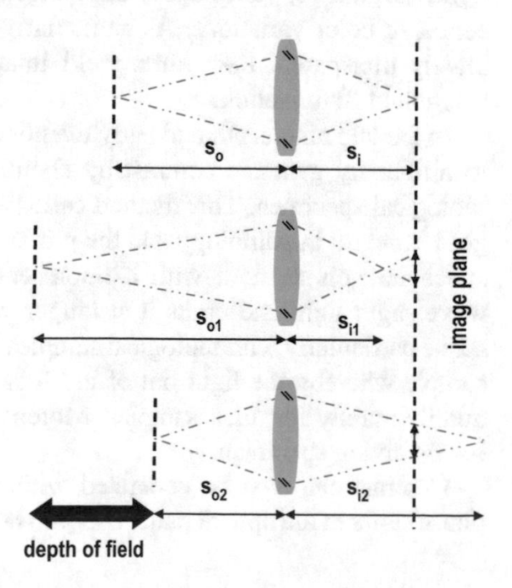

Fig. 5.16 Illustration of the depth of field in imaging

distance s_o is increased to s_{o1}. According to the Gaussian lens formula the distance s_i must now become smaller, namely s_{i1}. Then, the image point lies in front of the image plane. In the image plane a small circular spot of size d can be observed, the *circle of confusion*. A similar result is obtained when the object point is closer to the lens (s_{o2}). Then, the image point lies behind the image plane (s_{i2}) and again a circle of confusion can be observed in the image plane. This circle of confusion is the limiting factor for the vertical resolution as the observer or a detector cannot resolve it. Then, all object points within the range between s_{o1} and s_{o2} are perceived as "sharp" point. The distance $s_{o2} - s_{o1}$ defines the depth of field. It is this depth of field that prevents 3D imaging with a high vertical resolution. Other techniques of imaging or generation of evaluable signals must be used to establish highly resolved 3D imaging. These techniques will be discussed in the following Sect. 5.4.2 Confocal Microscopy.

The DoF can be improved by reducing the numerical aperture (making the aperture smaller or using lower numerical aperture objective lens) or by lowering the magnification for a given numerical aperture or by using a longer illumination wavelengths.

Sometimes, the phrase *depth of focus* is confusingly used for the depth of field. Depth of focus is however the distance over which the image plane can be displaced while a single fixed object plane remains in acceptably sharp focus, including tilt.

The classical microscope has undergone several improvements and extensions that are favorable in the main area of application of a classical microscope, the biological applications. The improvements are concerned with the *contrast* in the optical image. Contrast is defined as the difference in light intensity between the image and the adjacent background relative to the overall background intensity. To distinguish differences between the image and its background a minimum contrast value is needed. It amounts to 0.02 (2%) for the human eye but can have a different value for other detectors. Contrast is produced in the specimen by the absorption of light, brightness, reflectance, birefringence, light scattering, diffraction, fluorescence, or color variations. As particularly living or stained specimens are predominately transparent they often yield images with poor contrast when viewed in brightfield illumination.

In the late nineteenth and early twentieth centuries biologists developed methods to artificially generate contrast by chemically binding stains and fluorophores to biological specimen. This method called *fluorescence microscopy* is to this day the gold standard in pathology and the preferred imaging tool in cell biology. Here, the specimen gets marked with a fluorescent colorant. This colorant absorbs short wavelength light and emits it at longer wavelengths. The problem with this technique particularly with biological samples is that fluorescence occurs in the complete sample whereby the light out of the focal plane is blurred. This affects the image quality mainly for thick samples. Moreover, the contrast agents may be hazardous for the living specimen.

Contrast can also be generated without contrast agents introducing additional phase shifts in the optical path. This *phase contrast microscopy* was developed in the

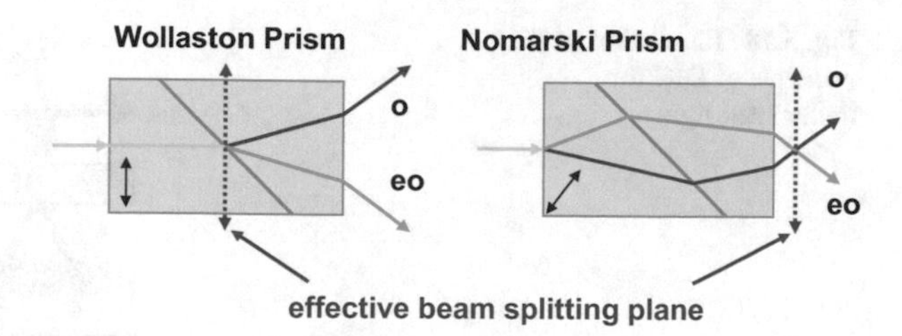

Fig. 5.17 Sketch of a Wollaston prism (left) and a Nomarski prism (right)

1930s by Frits Zernike [19]. He interpreted the microscope image as the result of interference between the incident light and light scattered by the specimen. As the scattered light is much weaker than the incident light the contrast must be poor. To increase the contrast of the image the contrast of the fringes in the image plane must therefore be increased. Zernike inserted a thin metallic filter called phase strip at the position of the back focal plane of the microscope objective that causes an additional quarter-wavelength shift between incident and scattered fields [20]. In addition, as the metallic phase strip exhibits significant absorption, it also balances the power of the two beams and further boosts the contrast. For this invention Zernike received 1953 the Nobel Prize in physics. When using circularly polarized light in microscopy instead of the linear polarized light of a single phase strip the contrast in the image can either be maintained independently of the azimuthal orientation of the examined specimen or it gets enhanced.

In the mid-1950s the idea of Georges Nomarski was to improve the contrast by using modified Wollaston prisms called *Nomarski prisms* to establish a *differential interference contrast* or abbreviated DIC [21, 22]. The difference in the Nomarski prism compared to the Wollaston prism is mainly the orientation of the optical axis in the birefringent crystal. For comparison, a Wollaston prism and a Nomarski prism are illustrated in Fig. 5.17.

In the Wollaston prism the incident light gets split into ordinary and extraordinary ray where the two birefringent crystal wedges are cemented together. This site is the effective beam splitting plane. These two rays vibrate perpendicular to each other. In the Nomarski prism the beam splitting occurs already when entering the first wedge due to the orientation of the optical axis in this wedge. Ordinary and extraordinary ray pass the first wedge and get redirected by the second wedge so that the separated rays finally appear as to come from an effective beam splitting plane outside the prism. Due to the different refractive index for the ordinary and the extraordinary ray and the longer paths through the prism both have a small path difference or shear at the effective beam splitting plane.

In Fig. 5.18 the working principle of DIC is explained for transmission microscopy.

The light from the light source first passes a polarizer located beneath the substage condenser. In between polarizer and condenser the plane-polarized light enters a Nomarski prism. It splits the entering beam into ordinary and extraordinary

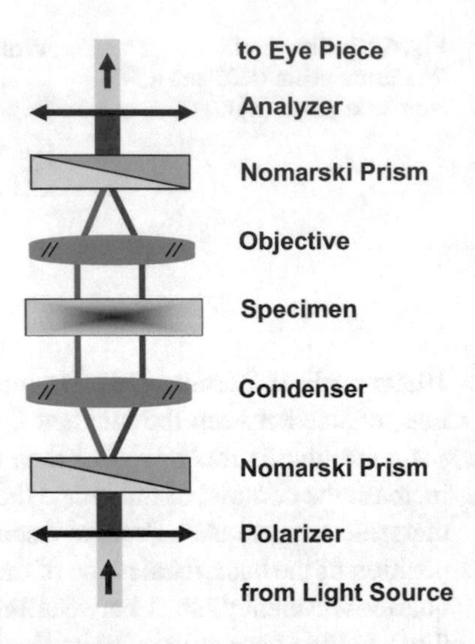

Fig. 5.18 Illustration of the principle of DIC for transmitted light

ray with slight path difference (shear) as described above. After passing then the condenser both rays enter and pass the specimen where their wave paths are altered in accordance to the specimen's properties (varying thickness, slopes, refractive indices). Next, both parallel rays are focused by the objective of the microscope in the back focal plane. There, a second Nomarski prism combines the rays again and removes the shear. As they have traversed the specimen the optical paths of the two beams are not of same length. Moreover, they still vibrate perpendicular to each other and cannot interfere. For interference the vibrations of the beams of different path length must be brought into the same plane and axis. This is accomplished by placing a second polarizer (analyzer) above the second Nomarski prism. The light then proceeds toward the eyepiece where it can be observed as differences in intensity and color. The design results in one side of a detail appearing bright (or possibly in color) while the other side appears darker (or another color). This shadow effect constitutes a pseudo three-dimensional appearance of the specimen.

C-DIC is the extension of DIC using circular polarized light. This makes the interference contrast invariant against the azimuth from the Nomarski prism. Then, the prism can be rotated fitting to the object properties. A further development of C-DIC is TIC (Total Interference Contrast) with a special Nomarski prism. It enables measurements on fine periodic surface structures but with a twin image.

An increase of the lateral resolution in classical microscopy can only be obtained by changing the wavelength λ. According to Eq. (5.4) it seems appropriate to use wavelengths in the UV. UV-microscopy indeed utilizes monochromatic UV light, often mercury at 248 nm, to obtain high resolution of up to 80 nm. On the other hand, special materials must be used for the optics in the microscope as most of the common glasses are not sufficiently transparent in this spectral range.

5.4.2 Confocal Microscopy

As mentioned in Sect. 5.4.1 the classical microscope suffers from the depth of field that prevents higher vertical resolution. Any object within the DoF will create a sharp image in the image plane. Hence, their heights cannot be resolved. Moreover, for a transparent thick sample thicker than the DoF, as to be often found in biological samples, a discrimination in depth along the sample thickness is not possible. Both problems are solved in the *confocal point sensor* and the *confocal microscope*.

The improvement is accomplished by a modified optical path in the microscope. It is illustrated in Fig. 5.19 for the confocal point sensor.

The light of a punctiform light source is first collimated and then reproduced in the focal plane of an objective at point P where it is reflected by the object. The reflected light is collected by the objective, gets deflected with a beam splitter, and is focused into a new focal plane that is the conjugated focal plane to the focal plane at the object's site. Here, the focused light passes a pinhole before its intensity is detected. But not only light from this object point arrives at the pinhole. Within the depth of field (DoF) of the optics also light from above-focal-planes and under-focal-planes gets collected by the objective and leads to a circle of confusion in the conjugated focal plane. That means that this light is widespread distributed in the conjugated focal plane. With the pinhole in front of the conjugated focal plane its brightness gets however strongly reduced so that predominately light from only the focal plane is just detected.

Unfortunately, even for an optical system that is corrected for diffraction a point is not imaged into a point but into a washed-out axial intensity distribution described by the *point-spread function* (PSF). The axial resolution is therefore determined by the PSF of the illumination PSF_{illum} in the focal plane of the objective convoluted with the PSF of the detection at the pinhole PSF_{det}

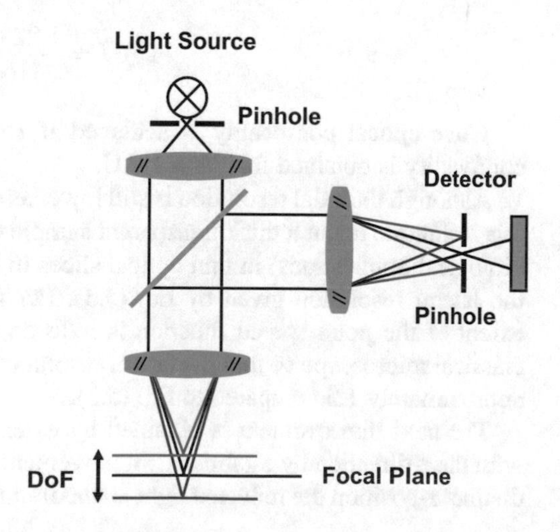

Fig. 5.19 Optical path in a confocal point sensor

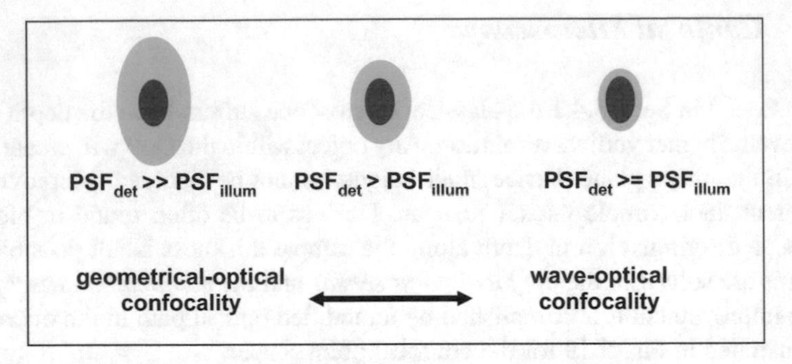

Fig. 5.20 Illustration of the PSF for geometrical-optical confocality and wave-optical confocality in a confocal point sensor or a confocal microscope

$$\mathrm{PSF} = \mathrm{PSF}_{\mathrm{det}} \otimes \mathrm{PSF}_{\mathrm{illum}} \tag{5.7}$$

Quantitatively the area at full width at half maximum (FWHM) of the corresponding PSF is considered. The ratio of PSF_{det} and PSF_{illum} defines two domains, the *geometrical-optical confocality* and the *wave-optical confocality* as illustrated in Fig. 5.20.

From this consideration of the PSFs the axial resolution follows as

$$\underbrace{\sqrt{\left(\frac{0.88\cdot\lambda}{1-\sqrt{1-\mathrm{NA}^2}}\right)^2+\left(\frac{\sqrt{2}\cdot\mathrm{PH}}{\mathrm{NA}}\right)^2}}_{\text{geometrical} - \text{optical}} \qquad \underbrace{\frac{0.64\cdot\lambda}{1-\sqrt{1-\mathrm{NA}^2}}}_{\text{wave} - \text{optical}} \tag{5.8}$$

and depends upon the numerical aperture *NA* of the used objective and the size *PH* of the pinhole in front of the detector. *PH* is usually measured in Airy units AU with

$$1\ \mathrm{AU} = \frac{1.22\cdot\lambda}{\mathrm{NA}} \tag{5.9}$$

Wave-optical confocality is achieved if $PH \leq 0.25$ AU, geometrical-optical confocality is obtained for $PH > 1$ AU.

Although the axial resolution is still lower as desired, it is obviously possible with this technique to cut a thick transparent sample (typically up to 100 μm thickness in biological applications) in thin optical slices in vertical direction while maintaining the lateral resolution given by Eq. (5.3). For fluorescence, the lateral (and axial) extent of the point spread function is reduced by about 30% compared to that in classical microscope so that the lateral resolution even becomes better by a factor of approximately 1.2 compared to Eq. (5.3).

The next improvement is obtained by extension to a scanning technique along with the *z*-direction by establishing a movement of the objective in *z*-direction. For a distinct *z*-position the reflected light is maximum if the workpiece or the sample is in

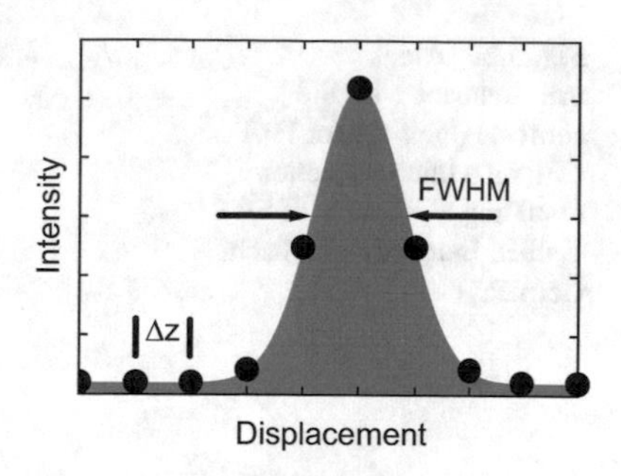

Fig. 5.21 Point-spread function in the conjugated focal plane scanned in steps of Δz of the displacement of the objective

the focal plane while for all other parts of the workpiece the reflected intensity is less. Trailing the intensity measured at the detector for varying z-position a curve is obtained like displayed in Fig. 5.21. It represents the point spread function PSF_{det} in the conjugated focal plane scanned in steps of Δz. Its FWHM is given again by Eq. (5.8). For reflective samples as in technical applications it is even smaller by about a factor 1.4 in the case of wave-optical confocality, i.e. it amounts then to

$$\mathrm{FWHM} = \frac{0.45 \cdot \lambda}{1 - \sqrt{1 - \mathrm{NA}^2}}. \tag{5.10}$$

Scanning with steps Δz clearly smaller than the DoF, it is thus circumvented and one can clearly assign the maximum of the intensity distribution to certain z-position of the objective. The vertical resolution is then in the order of the stepwidth Δz.

Commercially available confocal point sensors use an oscillating objective for axial scanning. Typical vertical scan ranges are in the order of 1000 μm with vertical resolution of 20 nm. The lateral resolution amounts to 1 μm for red laser light. Figure 5.22 shows an example of the surface topography of a micro injection molding measured with a confocal point sensor. The picture is courtesy of FRT GmbH, Bergisch Gladbach, Germany.

To analyze a larger part of the object in one shot the light spot must be scanned across the object area and the reflected light must be detected with a camera. This is realized in the confocal microscope. The result is a sectional view of the surface at position z_1 of the objective with some points being sharply imaged. Moving the objective to a second position z_2 a second sectional view is obtained with other points being sharply imaged. A complete topography measurement is compounded from 100 to 1000 of such sectional views or frames. From the maximum position in the intensity profiles of each pixel a 3D presentation of the topography is calculated. Similar to the confocal point sensor the vertical resolution of the microscope is significantly determined from the positioning accuracy of the positioner unit (linear

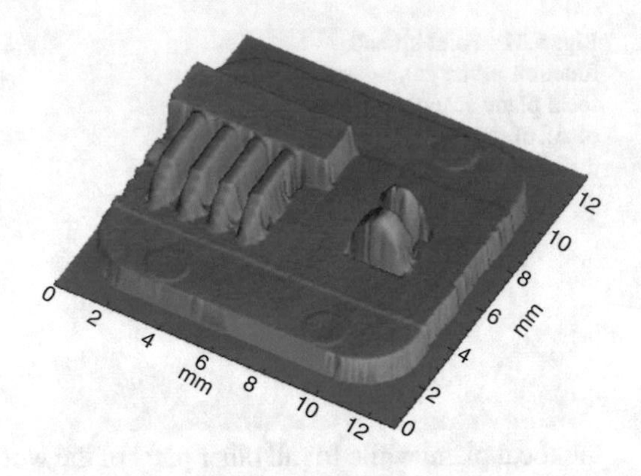

Fig. 5.22 Areal measurement with the confocal point sensor FRT CFP on a micro injection moulding. (Courtesy of FRT GmbH, Bergisch Gladbach, Germany)

stage, mostly yet a piezo actuator) and is in the order of the stepwidth Δz. A better resolution is obtained with advanced mathematical methods that are used to locate more precisely the intensity maximum. Real-time algorithms like a center of gravity algorithm as well as offline algorithms are in use. The advantage of offline algorithms is that they can deal with multiple peaks and optical artefacts as they store the entire series of confocal images in the computer's memory and calculate the 3D surface afterwards. Parabola fitting using some points around the maximum is one of the most utilized algorithms since it is fast, requires little data, and can provide resolution of up to 1/100 of the stepwidth Δz.

Field of curvature and other aberrations of the used optics are measured with a plane mirror with a planarity of $\lambda/40$. This measurement is provided as reference measurement. It is automatically subtracted from each further measurement. The remaining error then amounts to less than 10 nm.

For scanning the light spot across the specimen area three methods have been developed from which the first two methods are mostly used in commercial confocal microscopes.

The first method depicted in Fig. 5.23a utilizes either fast galvanometric rotating mirrors (at kHz frequencies) or MEMS mirrors in x- and y-direction that projects the parallel beam of the light source on a lens (scanning ocular) where it is focused. The resulting focus point constitutes the punctiform light source for the confocal imaging. Commonly, a laser is used as light source for which reason this kind of confocal microscope is called *laser scanning confocal microscope* (LSCM). The reflected light or the laser-induced fluorescence light (in biological applications) passing the pinhole in front of the detector gets collected with a photomultiplier. This is the basic concept of confocal microscopy. It was originally developed by Marvin Minsky in the mid-1950s with a patent in 1961 [23, 24]. His aim was to image neural networks in unstained preparations of brain tissue and was driven by the desire to image biological events at they occur in living systems.

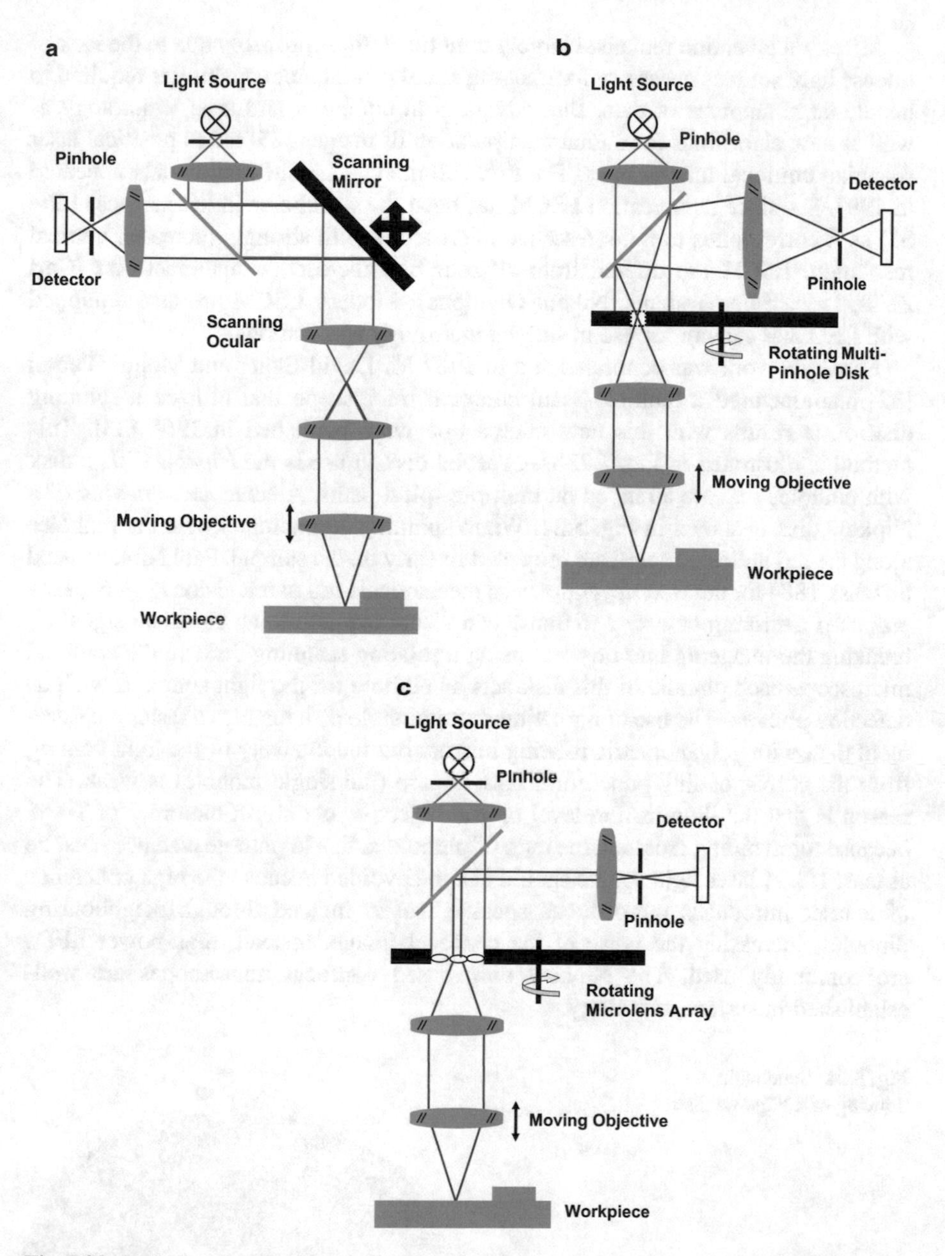

Fig. 5.23 Optical path of a confocal microscope (**a**) with scanning mirrors, (**b**) with rotating multi-pinhole disc (Nipkow disc), (**c**) with rotating array of microlenses

Minsky's invention remained largely unnoticed, most probably due to the lack of intense light sources necessary for imaging and the computer capabilities required to handle large amounts of data. But, advances in computer and laser technology as well as new algorithms for digital manipulation of images [25] led to practical laser scanning confocal microscopes. The first commercial LSCM instruments appeared in 1987. So, since its invention LSCM has been the workhorse in life sciences [26–31] as it corresponds to a fluorescence microscope with strongly increased vertical resolution. LSCM are offered from all four big microscope manufacturers (Carl Zeiss, Leica Microsystems, Nikon, Olympus). Modern LSCM are also equipped with LED and camera for use in surface metrology applications.

Minsky's work was continued and in 1967 M. David Egger and Mojmir Petran [32] manufactured a multiple-beam confocal microscope that utilized a spinning disk. First results with this new microscope were published in 1968 [33]. This method is illustrated in Fig. 5.23b. A special disk in use is the *Nipkow disk*, a disk with pinholes that are arranged on multiple spiral paths. A schematic drawing of a Nipkow disk is shown in Fig. 5.24. When spinning counterclockwise the pinholes along the red indicated spiral are projected linearly on the sample. Paul Nipkow used this disk 1884 for the *electric telescope*, a mechanical kind of television [34]. Nipkow proposed a striking new way to translate a visual image into an electrical signal by breaking the image up into tiny bits using a rotating scanning disk. In the confocal microscope each pinhole of this disk acts as pinhole for the light source as well as detection pinhole. The use of a rotating multi pinhole disk enables a faster measurement than with galvanometric rotating mirrors but the intensity of the light coming from the corresponding punctiform light source (the single pinhole) is weak. The reason is that the illumination level on the objective is only in the order of 1–9% because for avoiding crosstalk the ratio of pinhole radius to pinhole distance must be at least 1:3. A laser light source should also be avoided because the high coherence of a laser introduces out-of-focus speckle that is imaged through neighbouring pinholes, increasing the noise of the confocal image. Instead, high-power LEDs are commonly used. The Nipkow disk based confocal microscopes are well-established in surface metrology.

Fig. 5.24 Schematic drawing of a Nipkow disk

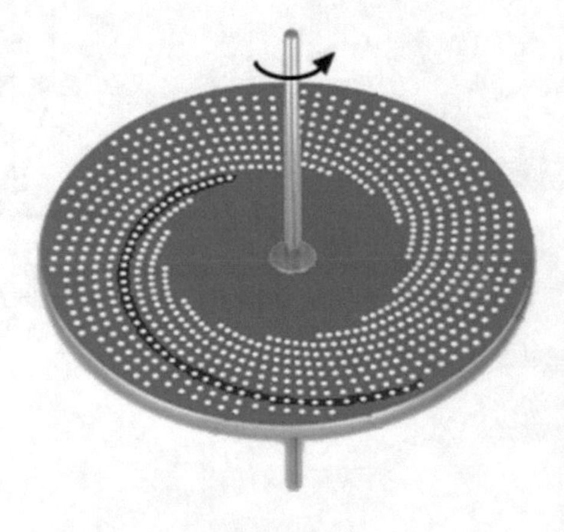

The third method shown in Fig. 5.23c uses a spinning disk with microlenses instead of a spinning pinhole disc [35, 36]. With such a disk the amount of light in the system can be increased by up to ten times. Then, the illumination level can almost achieve 100%.

Confocal microscopy has the advantage of imaging of smallest structures and steep edges with high *NA* and high vertical resolution. Vertical resolution for surface profiling is often defined as r.m.s. (root mean square) of the difference measurement of two smooth surfaces. The vertical resolution also determines the minimum step height that can be measured on a sample. It is further almost independent of the material of the surface. High measurement rates (about 5–10 s measurement time with evaluation inclusive) and high reproducibilities are common. To enable measuring areas larger than the field of view of the used objective many neighboring areas of same size are stitched with overlapping regions. Typical overlapping values are between 10% and 25% of the field. Sophisticated software algorithms use the information in the overlap regions to put the images together to a larger image. Nevertheless, stitching of multiple images can generate artifacts.

Modern confocal microscopes are completely integrated electronic systems where the optical microscope plays a central role in a configuration that consists of one or more electronic detectors, a computer (for image display, processing, output, and storage), and several illumination systems. An example for a modern, commercially available confocal microscope is shown in Fig. 5.25. The picture is courtesy of Olympus Europa SE & Co. KG, Hamburg, Germany.

Fig. 5.25 Laser Scanning Confocal Microscope OLS5000. (Courtesy of Olympus Europa SE & Co. KG, Hamburg, Germany)

This microscope can be used in the modes LSCM, reflection color bright field microscope, and laser/color differential interference contrast. It is equipped with a laser (405 nm wavelength) and a white light LED. Pictures are taken with a CMOS color camera and for laser confocal measurements with a two-channel photomultiplier tube. Instead of a galvanometric mirror a MEMS mirror from own development fastens the scanning process in the confocal mode.

Examples of measurements with a confocal microscope are shown in Fig. 5.26. The pictures are courtesy of FRT GmbH, Bergisch Gladbach, Germany. The picture at left shows a part of a microfluidics component with circular bumps recorded with a 20× objective. On the right side a laser inscription dot on a semiconductor wafer is

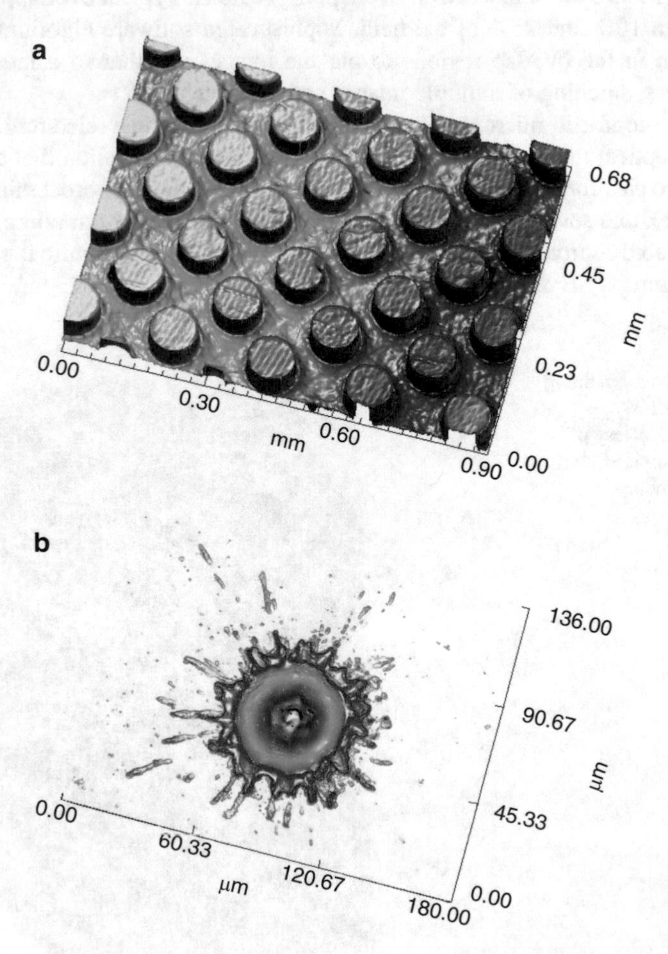

Fig. 5.26 Areal measurements with the confocal microscope FRT CFM: (**a**) part of a microfluidics component with circular bumps (20× objective), (**b**) laser inscription dot on a wafer (100× objective). (Courtesy of FRT GmbH, Bergisch Gladbach, Germany)

depicted recorded with a 100× objective. One can clearly recognize the throw-off of wafer material around the inscribed hole.

The LSCM particularly used in fluorescence microscopy has been further improved to get an increased lateral and vertical resolution. A dramatic improvement in the vertical resolution was obtained with the invention of *4π-microscopy* by Stefan Hell [37–39]. Two focused beams of the opposite objectives of two confocal microscopes superpose so that they enhance their fields in the common focus point. A second effect of this superposition is that the focal volume is remarkably reduced compared to a single confocal microscope. With this method Hell achieved in a decrease of the focal volume by a factor 3–5 compared with the confocal microscope of those days. Hell implemented with this kind of microscope the theoretical considerations of Thomas and Christoph Cremer in 1971 on the generation of an ideal 4π hologram for overcoming Abbe's diffraction limit [40, 41].

The *stimulated emission depletion* method or STED [42, 43] overcame first Abbe's diffraction limit for the lateral resolution although it also uses focused light. Resolutions of some few nanometers were obtained. The difference to classical optical fluorescence microscopy is that fluorescence molecules within the volume given by the diffraction limit are selectively switched "on" or "off" by stimulated emission The molecules in the volume of the focus diameter get excited to fluorescence with laser light of a certain wavelength. Then, extra light with a longer wavelength is used to deplete an excited molecule into the ground state or another non-fluorescent state. The trick is that this extra light has a zero intensity position in its spatial distribution. In addition, the fluorescence exciting light is as weak as only molecules in the vicinity of the zero position of the extra light become excited. This region is smaller than the region given by Abbe's diffraction limit. By this way the molecules that lead to images can be distinguished from the others with an improved spatial resolution of presently down to 20 nm. The use of other dark states beyond the ground state is realized in the RESOLFT method (*re*versible *s*aturable *o*ptical *fl*uorescent *t*ransitions) [44, 45] Complementary methods like PALM (*p*hoto-*a*ctivated *l*ocalization *m*icroscopy) [46], STORM (*st*ochastic *o*ptical *r*econstruction *m*icroscopy) [47], or SOFI (*s*uper-resolution *o*ptical *fl*uctuation *i*maging) [48] use the same on/off switching principle but switch on or off only one single molecule in the diffraction region. STED, RESOLFT, PALM, STORM, and SOFI are now routinely used in a growing number of laboratories to probe transparent matter at length scales of a tiny fraction of the wavelength of the imaging light. But there is a price to pay for gaining more spatial resolution: accurate sample preparations and optimized labels become critical. Calculations showed that the size of the spot and hence the lateral resolution follows a new rule

$$A = \frac{1.22 \cdot \lambda}{2 \cdot NA} \cdot \frac{1}{\sqrt{1 + \frac{I}{I_{sat}}}} \qquad (5.11)$$

where I is the intensity of the depleting light and I_{sat} is the characteristic saturation intensity for the specific molecules, a kind of threshold up which the fluorescence is prevented with a certain probability (50%). Increasing the ratio I/I_{sat} improves continuously the resolution. For $I = 0$, i.e. without stimulated emission, Abbe's result is obtained again.

5.4.3 *Focal Depth Variation*

The method of *focal depth variation* bases on a vertical scanning of the surface with a conventional microscope combined with a fast highly sophisticated mathematical algorithm. The method is also known as *Through-focus Scanning Optical Microscopy* (TSOM) or *Focus Variation* or *Shape from Focus*. The method is utilized in digital microscopes.

The principle of focal depth variation is illustrated in Fig. 5.27. The surface gets illuminated coaxially with modulated white light. The reflected light is used to obtain an image of the surface. Now the distance between surface and objective is

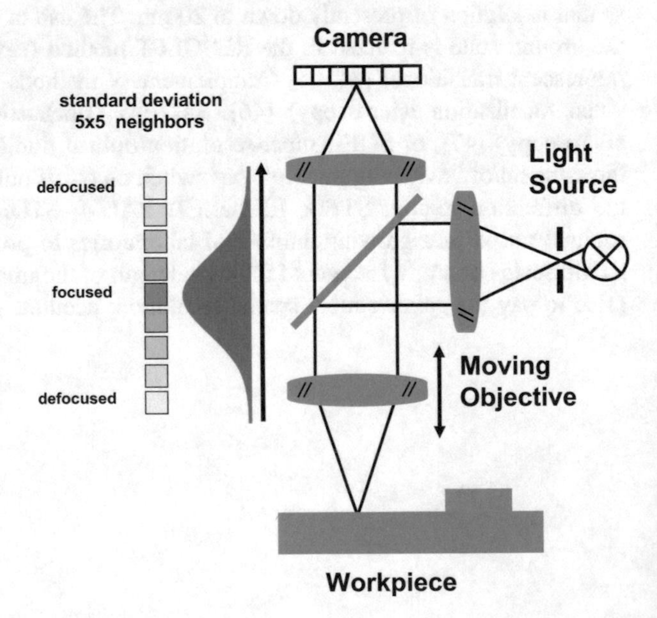

Fig. 5.27 Principle of focal depth variation

continuously varied (vertical scanning) whereby continuously digital images are taken. For each vertical position another part of the surface is in focus. However, due to the depth of field only a thin sheet of the measuring volume is exactly in focus. The task is to extract this sheet of best focus from the permanently collected data. As this change of focus is related to a change of contrast on the CCD sensor, the contrast can be analyzed to determine the best focus position. For example, one method is to calculate the standard deviation of the gray values in a small local quadratic region of M pixels (M being an odd number) around a certain pixel P. If the focus level is very low, i.e. the object is far from the focus plane the gray values of the M pixels are pretty similar. In consequence, the standard deviation is low. On the other hand, if the object is in the focus plane, the gray values in the M pixels differ stronger resulting in a large standard deviation. Plotting the results for the standard deviation versus the vertical position one gets an almost symmetric curve with a maximum as shown on the left side of Fig. 5.27. For determination of the maximum several methods are in use with increasing accuracy from top to bottom:

- Maximum point,
- Polynomial curve fitting, and
- Point spread function curve fitting.

This mathematics is accomplished using a highly sophisticated software that calculates the best focus sheet from the variation of the focus in the continuously recorded data. By this way the 3D surface is reconstructed. The focal depth variation is included in the ISO 25178 norm.

The focus variation method depends upon a sufficient roughness of the workpiece. It fails on highly reflecting samples and transparent specimen. The reason is that the focus must significantly change during the operation. For highly reflecting samples and transparent samples this condition is not fulfilled. Focal depth variation works well at surfaces with roughness values S_a larger than about 15 nm. In commercially available microscopes utilizing focal depth variation the lower limit for the roughness lies at $S_a = 50$ nm and $R_a = 80$ nm when using an objective with $100\times$ magnification. On the other hand, even inclination angles up to 90° can be evaluated. Vertical resolutions up to 10 nm and vertical uncertainties of up to 50 nm are achievable. Laterally, small field of views of 0.1×0.1 mm^2 as well as large field of views of 100×100 mm^2 can be achieved. The lateral resolution is again limited by the microscope resolution. The vertical scan height is limited by the working distance of the used objective and ranges between 3 and 22 mm. It is important that the scan range is a little higher than the vertical range of the sample. This is needed because focus variation uses a certain range before and after the maximum peak in the focus curve to calculate the depth value. The additional range is dependent on the depth of field. Ideally, adjustment of the scan range is automatically carried out by the instrument. The use of polarized light for illumination and the detection of the

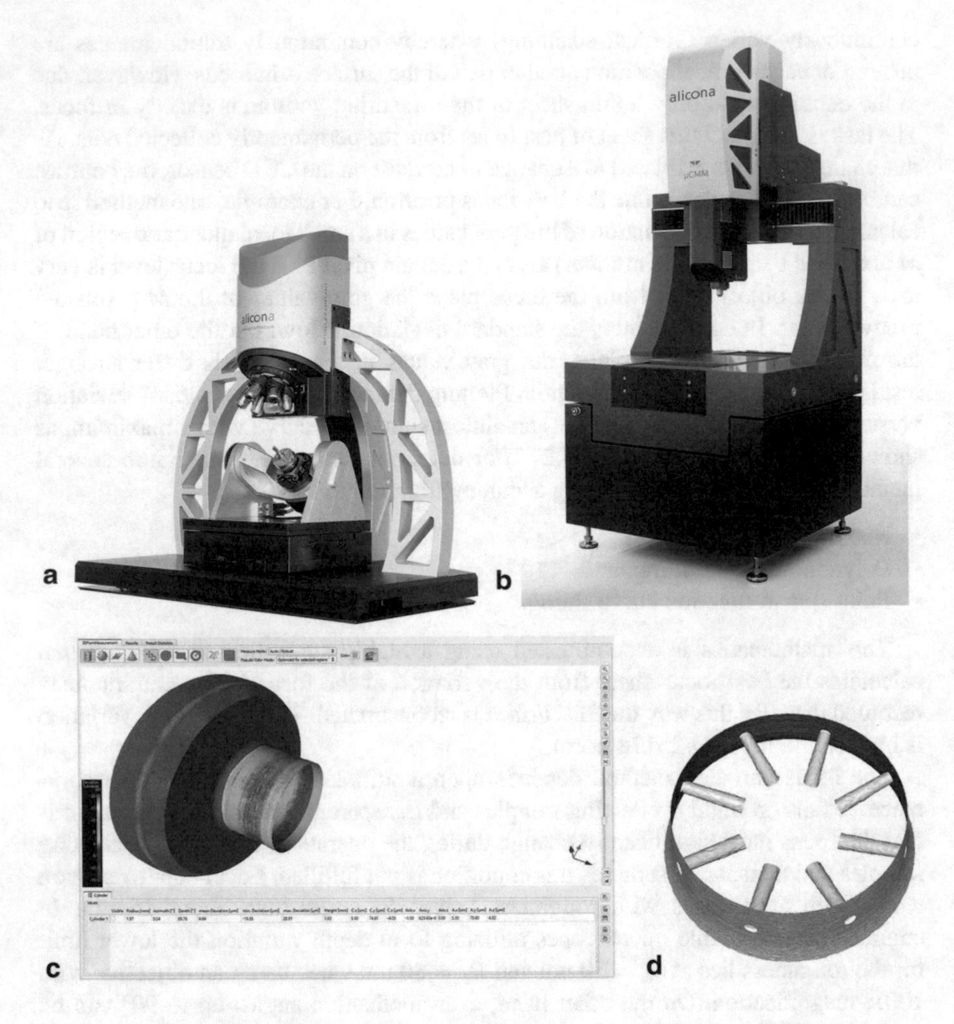

Fig. 5.28 (**a**) InfiniteFocus® and (**b**) μCMM utilizing focal depth variation, (**c**) Full form measurement based on Real3D technology of precision rotary component, (**d**) 3D hole measurement on the example of an injection valve. (Pictures are courtesy of Alicona Imaging GmbH, Raaba, Austria)

reflected light with an analyzer is beneficial when measuring metal samples with high reflectivity. Then, specular components that produce problematic highlights on the camera can be reduced.

Commercial measuring systems with focal depth variation are available for example from Alicona Imaging GmbH, Raaba, Austria. Representative pictures of measuring systems and obtained results are shown in Fig. 5.28. The pictures are courtesy of Alicona Imaging GmbH.

5.4.4 *Scanning Near-Field Optical Microscopy*

With the development of the optical near-field microscopy [49–54] shortly after the invention of the scanning tunneling microscope a scanning technique has been established that allows spatial resolution beyond the diffraction limit with optical means and at same time uses the versatile methods of optical spectroscopy. The idea to the *scanning near-field optical microscope* (SNOM) traces back to Synge [55]. He proposed to obtain a super-resolution beyond Abbe's diffraction limit when scanning over a surface with an opaque screen with minute aperture of 10 nm in diameter. The surface must be scanned at very proximate distances so that the light shining through this aperture can interact with the object. Synge also pointed to the challenges in fabricating this microscope. Although his proposal was simple and visionary it was far away beyond the technical capabilities of the time. Later, in 1972 Ash and Nicholls [56] demonstrated the near-field resolution of a sub-wavelength aperture microscope using microwaves. An overview on the theories on near-field optics can be found in [57].

Scanning near-field optical microscopes work in the near-field of a sample in contrast to conventional microscopes that work in the far-field. That means that the illumination of the sample and the detection of reflected or transmitted light are very close to the surface. The working principle of a SNOM is sketched in Fig. 5.29.

Usually, the light from a laser is focused on the surface through an aperture with a diameter smaller than the laser wavelength. This small aperture prevents normal propagation of the light through the aperture towards the surface in the lower part of the aperture. Instead, a so-called evanescent field is created with its amplitude decreasing exponentially in vertical direction. When the sample is now scanned at a small distance below the aperture, the evanescent field interacts with the surface and generates reflected light. The optical resolution of the reflected light is then limited only by the diameter of the aperture and not by the wavelength of the laser light. By this way this technique overcomes the classical lateral resolution limit of approximately half of the used wavelength. The optical resolution attainable is

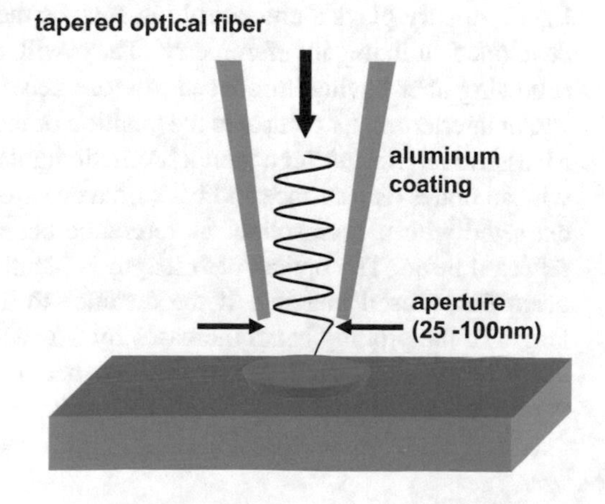

Fig. 5.29 Working principle of a scanning near-field optical microscope (SNOM)

typically in the range of 60–100 nm, values up to 5–10 nm are achievable under laboratory conditions. The optical image of the surface is generated by scanning the surface point-by-point and line-by-line. Commonly, tips of quartz are used. The quartz tips have typical aperture diameters of about 10 nm. A metal coating on the side wall ensures that the evanescent field is generated at the end of the tip. The tip is moved with piezo actuators very close to the surface (only several nanometers). The reflected light comes from a very small volume and enters the tip where it is directed to a photodetector. The metal coating on the side wall prevents loss of light after entrance in the tip. Instead of quartz tips also tips of silicon on a silicon cantilever are in use with light in the near infrared spectral region where silicon becomes transparent. The high refractive index of Si is sufficient to generate the evanescent field close to the end of the tip without metal coating. SNOM microscopes of WITec Wissenschaftliche Instrumente und Technologie GmbH, Ulm, Germany, are even equipped with unique, patented, high-quality micro-fabricated SNOM sensors consisting of a silicon cantilever with a hollow aluminum pyramid as a tip. The SNOM aperture is at the apex of the pyramid. The use of cantilevered tips is favorable because then the same precise probe-sample distance regulation as for AFM can be used. Otherwise, the probe-sample distance regulation relies on the detection of shear forces between the end of near-field probe and the sample. Shear force based systems allow simultaneous shear force and near-field imaging, including a transmission mode for transparent samples, a reflection mode for opaque samples, and a luminescence mode for additional characterization of samples.

Fields of application are investigation of near-field effects in microstructure metrology or the examination of the material composition of submicron structures. The latter is of big interest in semiconductor technology as well as in biology and medicine.

5.5 Interferometric Methods

Interferometry plays a crucial role in surface metrology. Many methods have been developed utilizing interferometry. They will be presented and discussed in the following after having introduced in more general principles of interferometry.

An interferometer compares the position or the surface structure of two objects. In a classical interferometer a monochromatic light source with wavelength λ is imaged with an optics on a surface and back on a detector. Additionally, a part of the light is extracted with a beamsplitter as reference beam, delayed, and interfered with the reflected beam. The optical path length is identical for reference beam and probing beam for focused imaging. If the distance to the surface increases by Δ the path length of the probing beam increases by 2Δ, while the path length of the reference beam keeps unchanged. When the two beams recombine, the observed intensity varies depending on the amplitude and phase of these beams

$$I(z,\phi) = I_{DC} + I_{AC} \cdot \cos\left(\frac{2\pi}{\lambda} \cdot z + \phi\right) \quad (5.12)$$

with the abbreviations $I_{DC} = I_1 + I_2$ and $I_{AC} = 2 \cdot \sqrt{I_1 \cdot I_2}$ and I_1, I_2 being the intensities of the reference beam and the sample beam.

Then, a continuous linear movement of the objective of the imaging optics results in a sinusoidal intensity variation (fringes) on the detector due to the interference. From the clear correlation between the sinusoidal intensity variation and the position of the objective the height of the specimen can be retrieved. However, due to the periodic intensity variation the axial position can only be determined modulo $\lambda/2$. A reconstruction of the complete surface topography is possible if the variation in height between two sampling points is less than $\lambda/4$. Traditionally, interferograms were measured by locating the center of a fringe and then tracing along the fringe. This method enables a fast determination of topographic variations of a workpiece in comparison to a reference plane. But rough surfaces cannot be measured with classical interferometry. Moreover, the setup of the measuring device must be pretty stable to prevent errors from vibrations.

As it is a difficult task to precisely locate the maxima or minima of the fringe intensity pattern, most optical-testing interferometers use phase shifting techniques because phase shifting is highly accurate and provides surface height measurements with Angstrom or sub-Angstrom resolution in good environments. *Phase shifting interferometry* (PSI) has the additional advantage to be insensitive to spatial variations of intensity, detector sensitivity, and fixed pattern noise. The earliest reference to PSI is believed to be Carré [58], however the development and demonstration of PSI began in the 1970s [59–61]. The key publication of Bruning et al. [61] was the incentive for various efforts in the field of AC interferometry.

In PSI the phase difference between the interfering beams is either changed in discrete steps by shifting the reference mirror a well-defined amount after taking an image or it is changed continuously at a constant rate while the detector is read out [62]. In doing so, multiple interferograms with different phase shifts ϕ are taken successively. This allows to retrieve exactly the position z from the phase of the cosine term in Eq. (5.12) by evaluating measured intensities.

A commonly exercised method is to shift the reference mirror by $\lambda/8$ to achieve in phase shifts of 0°, $\pi/2$, π, and $3\pi/2$ of the reference beam compared to the sample beam. Then, one obtains for each position *z* four equations

$$I_1 = I_{DC} + I_{AC} \cdot \cos\left(\frac{2\pi}{\lambda} \cdot z\right) \quad (5.13)$$

$$I_2 = I_{DC} - I_{AC} \cdot \sin\left(\frac{2\pi}{\lambda} \cdot z\right) \quad (5.14)$$

$$I_3 = I_{DC} - I_{AC} \cdot \cos\left(\frac{2\pi}{\lambda} \cdot z\right) \quad (5.15)$$

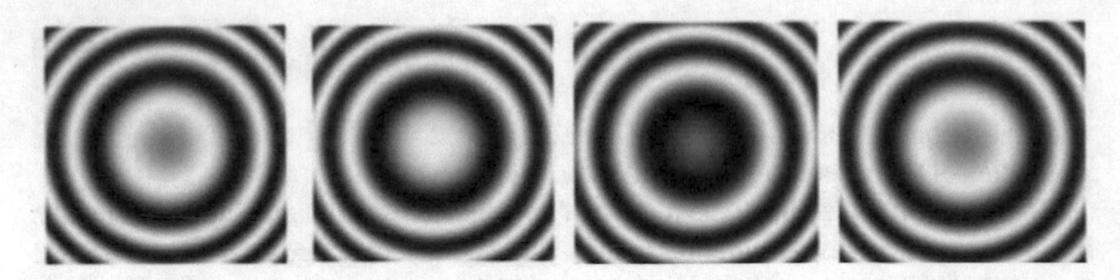

Fig. 5.30 Phase shifted interferograms

$$I_4 = I_{DC} + I_{AC} \cdot \sin\left(\frac{2\pi}{\lambda} \cdot z\right) \tag{5.16}$$

that can be resolved for the phase and hence for $z(x, y)$

$$\frac{2\pi}{\lambda} \cdot z = \tan^{-1}\left(\frac{I_4 - I_2}{I_1 - I_3}\right) \tag{5.17}$$

An example of corresponding phase shifted interferograms is shown in Fig. 5.30. The determination of the height z from Eq. (5.17) presumes that the shift between the four measurements is well-known and amounts to $\pi/2$. If it does not amount to $\pi/2$ but amounts to a fixed but unknown value another PSI algorithm must be used. One of them for $N = 4$ interferograms is the Carré algorithm [58]. It is assumed that the unknown phase shift amounts to 2α between each step, so that the values of the reference phase are

$$\frac{2\pi}{\lambda} \cdot z - 3\alpha, \ \frac{2\pi}{\lambda} \cdot z - \alpha, \ \frac{2\pi}{\lambda} \cdot z + \alpha, \ \frac{2\pi}{\lambda} \cdot z + 3\alpha. \tag{5.18}$$

The task is to resolve first for the unknown phase shift α and to use the result for resolving the phase. The phase shift α is obtained to

$$\alpha = \tan^{-1}\left(\sqrt{\frac{3 \cdot (I_2 - I_3) - (I_1 - I_4)}{(I_1 - I_4) + (I_2 - I_3)}}\right) \tag{5.19}$$

The result for the phase is then the Carré formula

$$\frac{2\pi}{\lambda} \cdot z = \tan^{-1}\left(\frac{\sqrt{[3 \cdot (I_2 - I_3) - (I_1 - I_4)] \cdot [(I_2 - I_3) + (I_1 - I_4)]}}{(I_2 + I_3) - (I_1 + I_4)}\right) \tag{5.20}$$

Equation (5.17) presumes that the shift between each of the four measurements is exactly $\pi/2$. If the phase shift exhibits linear variations from $\pi/2$ this phase shift error will result in a sinusoidal error in the reconstruction of the wavefront phase. The reconstruction can be improved by averaging two calculations of the phase with the second data set is shifted by $\pi/2$ compared to the first data set. In fact, this means, one takes one additional measurement I_5 and then calculates the phase from I_1, I_2, I_3, and I_4 and from I_2, I_3, I_4, and I_5 and takes an average of the two results. This is the algorithm developed by Schwider et al. [63, 64]:

$$\frac{2\pi}{\lambda} \cdot z = \tan^{-1}\left(\frac{2 \cdot (I_4 - I_2)}{I_1 - 2 \cdot I_3 + I_5}\right) \tag{5.21}$$

Hariharan et al. [65] developed another approach that is insensitive to the phase shift error. They used five measurements and initially assumed a linear shift α between the measurements: -2α, $-\alpha$, 0, α, and 2α. They showed that the reconstruction of the phase becomes insensitive to phase shift calibration errors when $\alpha = \pi/2$. Then, the reconstructed phase is obtained from the five measurements as

$$\frac{2\pi}{\lambda} \cdot z = \tan^{-1}\left(\frac{2 \cdot (I_2 - I_4)}{2 \cdot I_3 - I_5 - I_1}\right) \tag{5.22}$$

which is quite similar to the algorithm of Schwider et al. in Eq. (5.21).

The arcus tangens function in all above equations is restricted on the interval $[-\pi, \pi]$ by what the obtained phase profile is wrapped to this interval and the obtained phase profile contains one or more 2π jumps. It can be unwrapped using a suited unwrapping algorithm. The basic phase unwrapping process can be explained by the following steps.

1. Start with the second point from left in the wrapped phase profile.
2. Calculate the difference between the current point and its directly adjacent left-hand neighbor.
3. If the difference between the two is larger than $+\pi$, then subtract 2π from this point and also from all the points to the right of it.
4. If the difference between the two is smaller than $-\pi$, then add 2π to this point and also to all the points to the right of it.
5. Have all points been processed? If *No* then go back to step 2. If *Yes* then stop.

If desired, the intensity data can also be evaluated to determine the ratio I_{AC}/I_{DC}:

$$\frac{I_{AC}}{I_{DC}} = \frac{2\sqrt{(I_4 - I_2)^2 + (I_1 - I_3)^2}}{I_1 + I_2 + I_3 + I_4} \tag{5.23}$$

for the four measurements algorithm and

$$\frac{I_{AC}}{I_{DC}} = \frac{3\sqrt{4 \cdot (I_4 - I_2)^2 + (I_1 + I_5 - 2 \cdot I_3)^2}}{I_1 + I_2 + I_3 + I_4 + I_5} \tag{5.24}$$

for the five measurements algorithms.

5.5.1 Interferometric Form Inspection

Since interferometers are pretty sensitive on form deviations, the main field of applications are form deviations of flat samples and the quality control of spherical and aspherical lenses. The corresponding interferometric methods are described in the following two subsections.

5.5.1.1 Form Inspection of Planar Surfaces

For form inspection of planar surfaces most often Twyman-Green interferometers or Fizeau interferometers are used. The setups are illustrated in Fig. 5.31.

The Twyman-Green interferometer uses a punctiform light source and lenses in the reference and object arm to generate a collimated beam in front of the reference mirror and the object. For phase shifting the reference mirror can be moved. In the Fizeau interferometer the light of the punctiform light source is collimated by a lens and projected on the object. In between the lens and the measuring object a transparent flat is positioned that serves on the one hand as reference mirror and on the other hand as phase shifter. Fizeau interferometers have the advantage that a perturbation in the beam path modifies the reference wave and the object wave in the same way and can therefore be dropped off numerically.

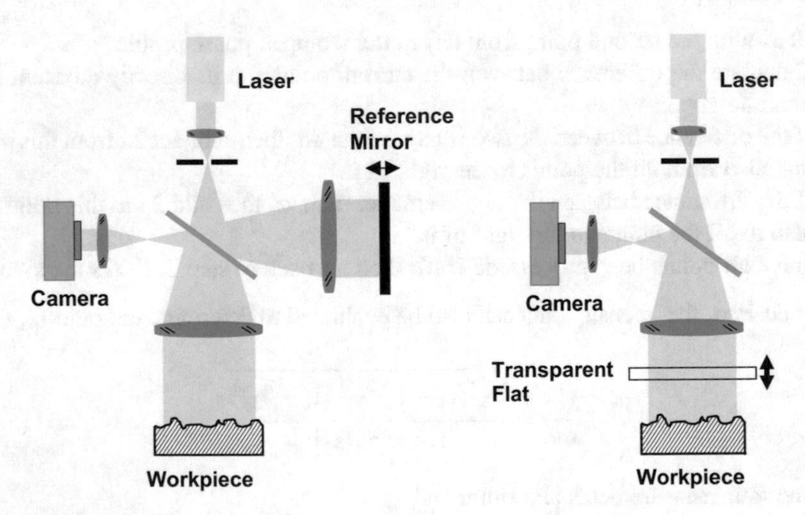

Fig. 5.31 Setups for form inspection with a Twyman-Green interferometer (left) and a Fizeau interferometer (right)

5.5.1.2 Form Inspection of Spherical, Aspherical and Freeform Surfaces

The use of aspherical and freeform surfaces becomes more and more important in the design of modern optical systems. Hence, the testing of such surfaces is an important task. For form inspection of spherical, aspherical, and freeform surfaces also Twyman-Green interferometers or Fizeau interferometers are used but with slightly modified setups compared to the form inspection of planar surfaces. They are illustrated in Fig. 5.32.

The setups of both interferometers are modified so that an additional lens in front of the object concentrates the light on an area of the curved surface of the object. Strong deviations between reference and object wave are usually compensated by a so-called null lens, a lens in form of a meniscus with both refractive interfaces having the same curvature (same radius of curvature of same sign) that is inserted in the beam path at a certain position. For optics with aspheric surfaces the "null lens" becomes more difficult. Yet, with the help of diffractive optical elements it is possible to achieve in a compensation of strong deviations. In practice, so-called *computer-generated holograms* (CGH) are made that generate a wavefront that is adjusted to the target surface of the specimen [66–74]. For the CGH the interferogram of the perfect aspheric wavefront with the reference beam is computed and then printed onto a mask for subsequent illumination. In both setups a possible interferogram from the light reflected at the rear side of the transparent object gets masked in front of the detector.

Measurements often take long time or they are less flexible. Further, they are expensive because of the production costs of the CGHs.

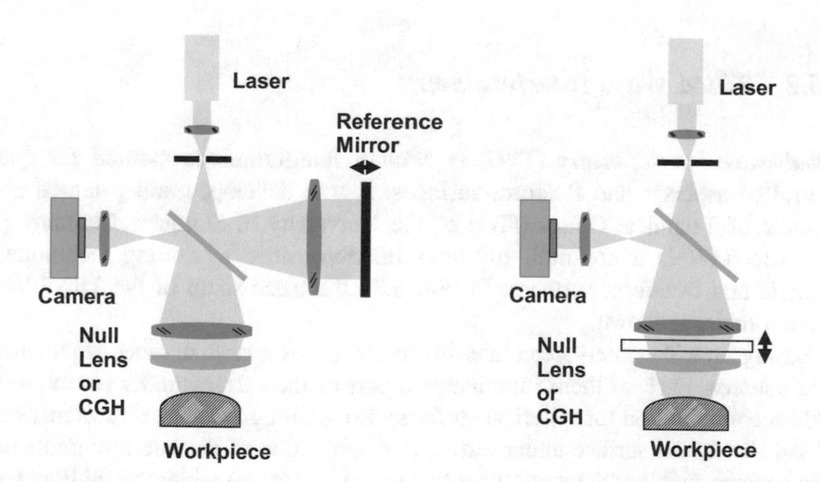

Fig. 5.32 Setups for form inspection of spheres, aspheres, and freeforms with a Twyman-Green interferometer (left) and a Fizeau interferometer (right)

For evaluation the measured wavefront is decomposed into a linear combination of *Zernike polynomials* (after the optical physicist Frits Zernike [75, 76]).

These polynomials are orthogonal on the unit disk. There exist even polynomials and odd polynomials with the definitions

$$\mathrm{Z_n^m}(\rho, \varphi) = \mathrm{R_n^m}(\rho) \cdot \cos(\mathrm{m}\varphi) \tag{5.25}$$

for the even polynomials and

$$\mathrm{Z_n^{-m}}(\rho, \varphi) = \mathrm{R_n^m}(\rho) \cdot \sin(\mathrm{m}\varphi) \tag{5.26}$$

for the odd polynomials. The angle φ is the azimuthal angle and m and n are nonnegative integer numbers with $n \geq m$. ρ is the normalized radial distance $0 \leq \rho \leq 1$. The functions $R_n^m(\rho)$ are the radial polynomials with definition as.

$$R_n^m(\rho) = \sum_{k=0}^{(n-m)/2} \frac{(-1)^k (n-k)!}{k! \cdot ((n+m)/2 - k)! \cdot ((n-m)/2 - k)!} \cdot \rho^{n-2k} \tag{5.27}$$

if (*n-m*) is even. For (*n-m*) being an odd number it is $R_n^m(\rho) = 0$.

The Zernike polynomials describe typical optical properties and errors of a lens or a lens system as, e.g. defocus, coma, or astigmatism. The polynomial decomposition gives a numerical representation of any kind of aberration of the sample. A graphical representation and their meaning of the first 21 polynomials is given in Fig. 5.33.

5.5.2 Tilted Wave Interferometry

Tilted wave interferometry (TWI) is another interferometric method for quality control of aspheric and freeform surfaces. It was developed and patented at the Institute of Technical Optics (ITO) of the University of Stuttgart, Germany [77–83]. The TWI is a non-null, full-field interferometric measuring technique for aspheric and free-form surfaces. In Fig. 5.34 the basic setup of the Tilted-Wave-Interferometer is shown.

Basing on a Twyman-Green interferometer it uses a large number of punctiform light sources. Each of them illuminates a part of the surface under test close to a nulltest configuration for spherical surfaces. So, each light source allows measuring a local part of the surface under test. The combination of the interferograms of all light sources enables the measurement of the entire surface under test without CGH. To avoid interference between neighboring sources the measurement is divided into four steps, with only every fourth light source enabled in each step. The switching between the sources is realized by a simple aperture array that is moved in front of the microlenses and blocks every second source in each row and column. The result

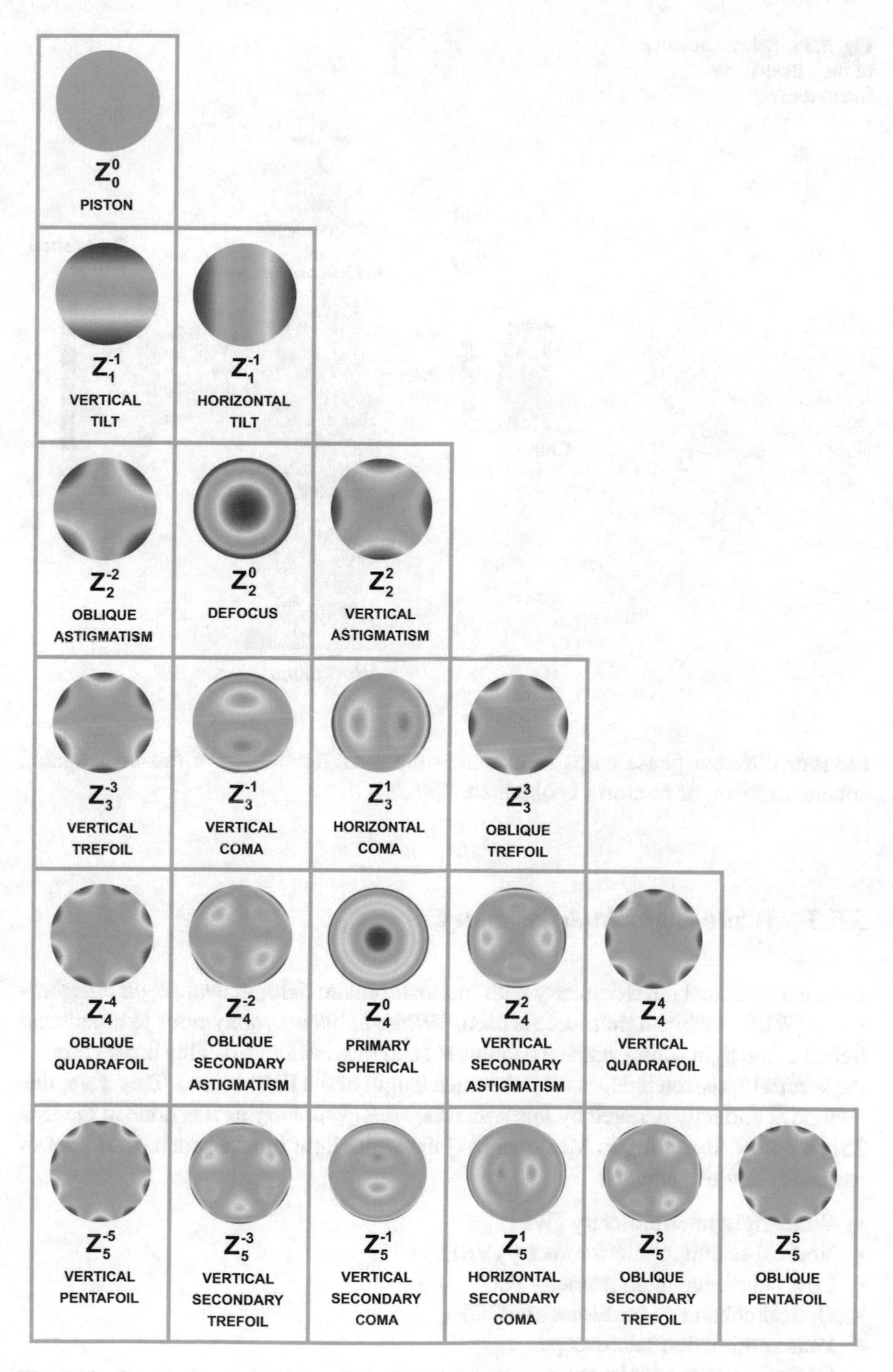

Fig. 5.33 Graphical representation of the first 21 Zernike polynomials and their meaning

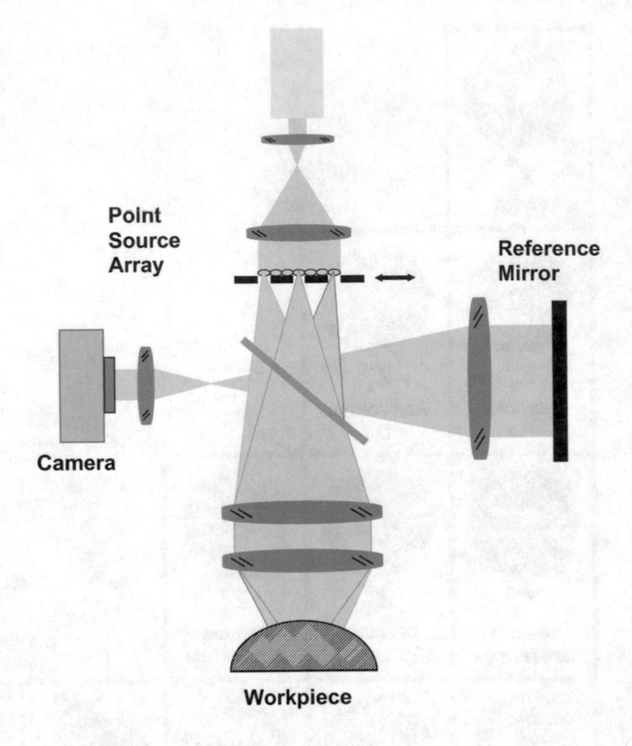

Fig. 5.34 Schematic setup of the Tilted-Wave-Interferometer

are four different phase maps from which the form deviations of the investigated sphere, asphere, or freeform is obtained.

5.5.3 White Light Interferometry

Unlike in classical interferometry with monochromatic light, in *white light interferometry* (WLI)) a white light source is used. "White light" is synonymous to broadband light, i.e. the light source emits a continous band of wavelengths. This broadening of the spectral emission reduces the coherence length of the light source. Therefore, this method is correctly denoted by *low coherence interferometry* as it is done in the ISO 25178 norm. According to Malacara [84] the white light interferometry has a lot of names. A few of them are:

- White light interferometry (WLI),
- Vertical scanning interferometry (VSI),
- Low coherence interferometry (LCI),
- Optical coherence profilometry (OCP),
- Phase correlation microscopy,
- Optical coherence microscopy,

- Interference microscope,
- Coherence radar.

In practice, the term white light interferometry (WLI) is most often used by manufacturers of such devices and is accepted by the practitioners.

The imaging system is comparable to that of a confocal microscope except of that the optical path includes a reference path for a part of the illuminating light. Similar to the confocal microscope the height information is obtained by vertical scanning, i.e. stepwise moving of the objective by Δz and taking an image at each step. The images (frames) then are combined to the final image of the object. With WLI height resolutions of up to 0.1 nm are obtainable while the lateral resolution is again limited by the diffraction limit. The height resolution is defined as r.m.s. of the difference to a measurement on a smooth sample.

The short coherence length of the light source results in spatially restricted interference patterns where the magnitude decreases from the center to the left and right similar to a natural wave. Then, the trailing of the intensity measured at one pixel of the camera results in a curve as it is displayed in Fig. 5.35.

The intensity correlogram yields a definite height information for each pixel from which the complete topography can be reconstructed. The correlogram consists of a periodic signal modulated by a Gaussian envelope.

$$I(z) = I_{DC} + I_{AC} \cdot \cos\left(\frac{4\pi}{\lambda_0} \cdot (z - z_0) + \phi\right) \cdot \exp\left(-2 \cdot \left(\frac{z - z_0}{l_c}\right)^2\right) \quad (5.28)$$

The width of the correlogram amounts to about three coherence lengths. To determine the topography of the surface the correlogram of each pixel must be evaluated. The task is to find the best focus position since it corresponds to the actual height $z(x, y)$ with zero optical path difference. To find this best focus position with a better resolution than the stepwidth Δz in the measurement different algorithms can be applied.

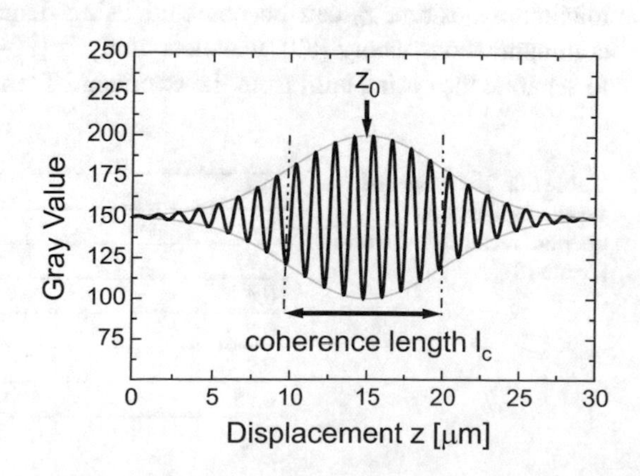

Fig. 5.35 Intensity correlogram at one pixel for a displacement of the objective of 30 μm

One method was developed by Larkin [85] and Sandoz et al. [86] called *white-light phase-shifting interferometry* (WLPSI). The first step is to locate the coherence peak position z_O where the envelope is maximum as position of the largest modulation contrast. For this the contrast M_n at the n-th frame (the position $z = n{\cdot}\Delta z$) is calculated from five neighboring frames in the correlogram via.

$$M_n^2 = (I_{n-1} - I_{n+1})^2 - (I_{n-2} - I_n) \cdot (I_n - I_{n+2}) \tag{5.29}$$

The maximal M_n yields the position z_O of the coherence peak with frame number m ($z_O = m{\cdot}\Delta z$). It does not generally coincide with the best focus position but is shifted by a phase shift $\Delta\Omega$. The next step is therefore to determine $\Delta\Omega$ from five neighboring frames around the coherence peak position, applying a five point PSI algorithm. Hence, WLPSI works well if $\Delta z = \lambda_0/8$. $\Delta\Omega$ is then obtained from

$$\tan(\Delta\Omega) = \frac{\sqrt{4 \cdot (I_{m-1} - I_{m+1})^2 - (I_{m-2} - I_{m+2})^2}}{2 \cdot I_m - I_{m+2} - I_{m-2}} \tag{5.30}$$

and the best focus position is

$$z_{best\ \ focus} = m \cdot \Delta z + f(NA) \cdot \frac{\lambda_0}{4\pi} \cdot \Delta\Omega. \tag{5.31}$$

where *f(NA)* is the so-called *numerical-aperture factor*. The *NA* of an interferometric microscope objective can affect the fringe spacing and thus the surface heights measured with that objective [87, 88]. According to Sheppard and Larkin [88] it is

$$f(NA) = \frac{2}{3} \cdot \frac{1 - \cos^3(\alpha)}{1 - \cos^2(\alpha)} \qquad \text{with} \quad \alpha = \sin^{-1}(NA) \tag{5.32}$$

Experimental results for *f(NA)* from Creath [87] are given in Table 5.2.

Instead of using the modulation contrast to find the coherence peak position, the maximum position z_O can be obtained using demodulation techniques based on communications theory [89], wavelets analysis [90–93], or other methods [94, 95] to separate the cosine term from the exponential term.

Table 5.2 Experimental results for the numerical-aperture factor f(NA) from Creath [87]

NA	f(NA)
0.1	1.003
0.25	1.007
0.4	1.024
0.5	1.036
0.9	1.215
0.95	1.228

Preferably, first the dc-component is removed from the signal. Then, the new intensity values are squared. The advantage of these squared intensities becomes obvious when considering the square of the intensity given in Eq. (5.28) after subtraction of the dc-term I_{DC}:

$$I_{corr}^2(z) = I_{AC}^2 \cdot \cos^2\left(\frac{4\pi}{\lambda_0} \cdot (z - z_0) + \phi\right) \cdot \exp\left(-4 \cdot \left(\frac{z - z_0}{l_c}\right)^2\right). \quad (5.33)$$

According to the addition theorems of the trigonometric functions it is $\cos^2(\psi) = 0.5 \cdot (1 + \cos(2\psi))$. This relation allows for rewriting Eq. (5.33) into

$$I_{corr}^2(z) = \frac{1}{2} I_{AC}^2 \cdot \cos\left(\frac{8\pi}{\lambda_0} \cdot (z - z_0) + 2\phi\right) \cdot \exp\left(-4 \cdot \left(\frac{z - z_0}{l_c}\right)^2\right) + \frac{1}{2} I_{AC}^2 \cdot \exp\left(-4 \cdot \left(\frac{z - z_0}{l_c}\right)^2\right) \quad (5.34)$$

The big advantage is that the envelope function is contained in a separate nonoscillating term. Then, a Fourier transform with a successive low pass filtering and inverse Fourier transform of the filtered data yields simply the (square of the) envelope function with maximum $z_0 = m \cdot \Delta z$. The best focus position can be determined again with a PSI algorithm. By this way, resolutions in the order of 0.1 nm are obtainable.

If Δz does not match the condition $\Delta z = \lambda_0/8$ exactly WLPSI cannot be applied. Nevertheless, good results for the height $z(x, y)$ and the topography are obtained by evaluating the centroid or center of gravity (COG) of the envelope. Preferably, again the dc-component is removed from the signal and the dc-corrected data are squared. One method is then to smooth the data with a smoothing filter, for example with a sliding average, to find the maximum of the smoothed data and to calculate the COG within an interval of $\pm N$ frames around this maximum. Another approach is to find the maximum of the squared but not smoothed data and to apply again Fourier transform and low pass filtering in the Fourier space to retrieve the envelope. Afterwards, a COG algorithm is applied on the envelope within an interval of $\pm N$ frames around the determined maximum. Both methods are robust and fast. The accuracy in determining z_0 depends upon the coherence length, the step width in the vertical scanning, vibrations, the contrast, and the surface roughness of the workpiece. Resolutions of 1 nm are obtainable with high aperture objectives.

As the COG is in principle a weighted average its value depends upon the used weights. Customarily, the frame positions $z_n = n \cdot \Delta z$ are weighted with the amplitudes I_n of the signal

$$\mathrm{COG} = \frac{\sum_{n} z_n \cdot I_n}{\sum_{n} I_n} \tag{5.35}$$

Another common algorithm calculates the square of the amplitude derivatives [96, 97].

$$\mathrm{COG} = \frac{\sum_{n} z_n \cdot (I_n - I_{n-1})^2}{\sum_{n} (I_n - I_{n-1})^2} \tag{5.36}$$

The white light interferometer started in the late 1980s and in the early 1990s [98–103]. It bases on the patent of *Mirau* [104] for the rugomètre – a roughness measuring tool. This interferometer works like a microscope with an objective where the reference beam is established by a semitransparent plate where half of the incident light gets reflected on a small mirror in the center of the objective and then further reflected to the camera by the same plate. The transmitted light hits the surface of the workpiece and gets reflected there. A sketch of a Mirau interferometer is shown in Fig. 5.36.

The interference of both beams in each camera pixel yields the correlograms when moving the objective in *z*-direction. A precise track of the objective over the whole measuring range is obtained with a piezoelectric drive. Regulated piezo systems enable up to 400 μm measuring range. Today, the Mirau-type interferometer is the most popular configuration for white light interferometry.

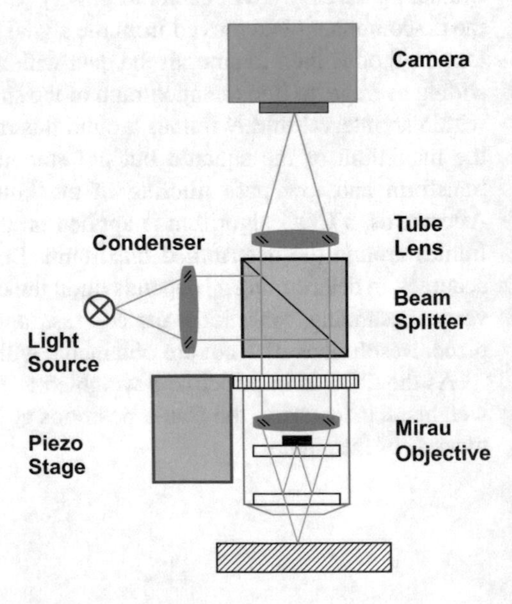

Fig. 5.36 Sketch of the Mirau-type white light interferometer

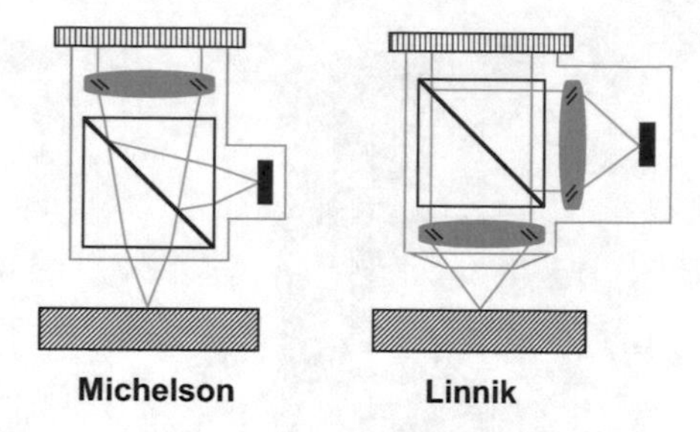

Fig. 5.37 Sketch of an objective with integrated Michelson interferometer (left) and of a Linnik objective (right)

Mirau objectives are available with numerical apertures between 0.3 and 0.6. For lower apertures (and low magnification) this type of objective is not well-suited. Instead, objectives with an integrated small Michelson interferometer are then used. Such an objective is sketched in Fig. 5.37 at the left side.

Although the Mirau and Michelson are the most common interference microscope objectives, they are not the only designs. Some high-magnification systems that demand larger working distance than the Mirau are served by the *Linnik* objective (see Fig. 5.37, right), constructed from two matched microscope objectives, one for the reference and one for the measurement path [105]. A disadvantage is the laborious adjustment. In practice the Michelson is the preferred geometry for low magnifications of 10× and lower. The Mirau is common for 20× to 100×, while the Linnik is sometimes used at 50× and 100×.

Similar to a confocal microscope the field of curvature and other aberrations of the used optics are measured with a plane mirror with a planarity of $\lambda/40$ and provided as reference. This reference is automatically subtracted from each measurement.

Problems arise with optical material properties of the specimen. The intensity correlogram in Fig. 5.35 is the ideal case which is obtained if both interfering waves are of same form and differ maximum in their intensity. Deviations of the form occur if the reference mirror and the specimen have different wavelength dependences of the reflectivity. This is the usual case. It can affect the correlogram threefold [106]:

- A shift of the complete correlogram,
- A shift of the maximum of the envelope, and
- A frequency change in the fringes of the correlogram.

These shifts will not be larger than about 40 nm but this is the error in height determination of surface structures if these shifts and distortions are not taken into account. Similar problems may arise with multiple reflections. They may happen in workpieces with deep holes where the incident beam gets reflected between the walls of the hole.

If the wavelength dependence of the reflectivity even leads to strong changes of the reflectivity within the bandwidth of the used light source distortions of the correlogram will result. Critical materials for which such distortions may occur are gold and gold alloys, copper and copper alloys, and titanium nitride.

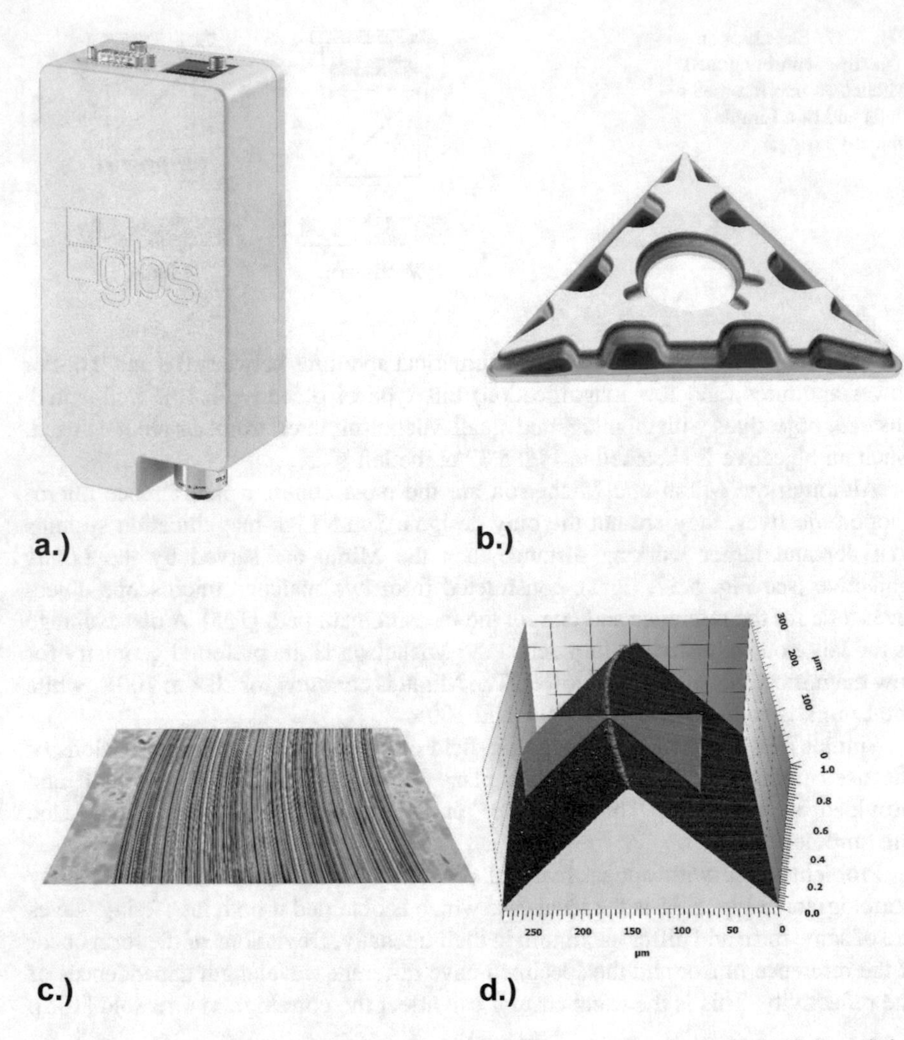

Fig. 5.38 (**a**) Smart WLI for industrial applications, (**b**) 3D topography of an insert (**c**) High resolution roughness measurement on a roughness standard, (**d**) 3D topography of a cutting edge. (Pictures are courtesy of Gesellschaft für Bild- und Signalverarbeitung (GBS) mbH, Ilmenau, Germany)

A commercial compact Mirau WLI is shown in Fig. 5.38 together with three exemplaric measurement results on the surface topography. The pictures are courtesy of Gesellschaft für Bild- und Signalverarbeitung (GBS) mbH, Ilmenau, Germany.

In another setup of a white light interferometer a Michelson interferometer is combined with a telecentric objective. The Michelson interferometer establishes an almost parallel probe and reference beam. It is arranged in front of a telecentric objective that images the interference pattern of the reflected light of each object point with the reference beam on a camera. The principal setup of this interferometer is shown in Fig. 5.39.

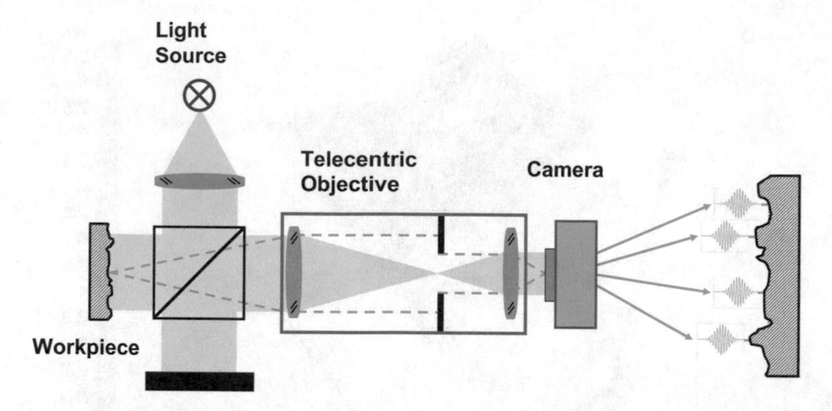

Fig. 5.39 Principal setup of a telecentric white light interferometer with telecentric objective

The lateral resolution of this telecentric WLI is clearly reduced because of the almost parallel illumination of the specimen. It is more or less given as the size of the field of view divided by the number of camera pixels in x- and y-direction. On the other hand here the field of view can amount up to $50 \times 50\ \text{mm}^2$. Even larger field of views can be realized but with vast efforts. With such large fields of view and collimated illumination the telecentric WLI are well-suited for measurements of planarity, evenness, or nanotopography. An example for nanotopography is given in [107]. Such telecentric WLI are available, e.g., from ISRA VISION AG, Darmstadt, Germany or from Polytec GmbH, Waldbronn, Germany. As for this type of WLI the complete interferometer gets moved with a linear stage the achievable stepwidth often does not match $\lambda_0/8$ for which phase-shifting algorithms cannot really be applied.

In Fig. 5.40 two exemplaric measurements with this parallel beam WLI are shown. The pictures are courtesy of ISRA VISION AG, Darmstadt, Germany. The first picture is the result of a roughness measurement inside an internal cone of about 8.7 mm in height. The second picture shows the result of an evenness measurement on a sealing with 143 mm diameter.

The use of any kind of WLI requires a tilt correction of the workpiece prior to measuring. This can be done for plane surfaces but for curved surfaces the interferences of same inclination cannot be completely compensated. Another problem which is common to all kinds of interferometers is the sensitivity to vibrations. Active shock absorbing tables are recommended when using white light interferometers.

A more challenging case for any WLI is a workpiece that is coated with a partially transparent thin film. Then, the correlogram contains three readily identifiable signal components, corresponding to the top surface, the substrate surface, and the layer thickness interference. They can clearly be recognized in Fig. 5.41. Be aware that the magnitude of the bottom surface correlogram may be higher than that of the top surface in contrast to the illustrated correlogram in Fig. 5.41. This happens when the reflection at the film-substrate interface is higher than at the interface film- air, for example for a transparent layer on silicon substrate.

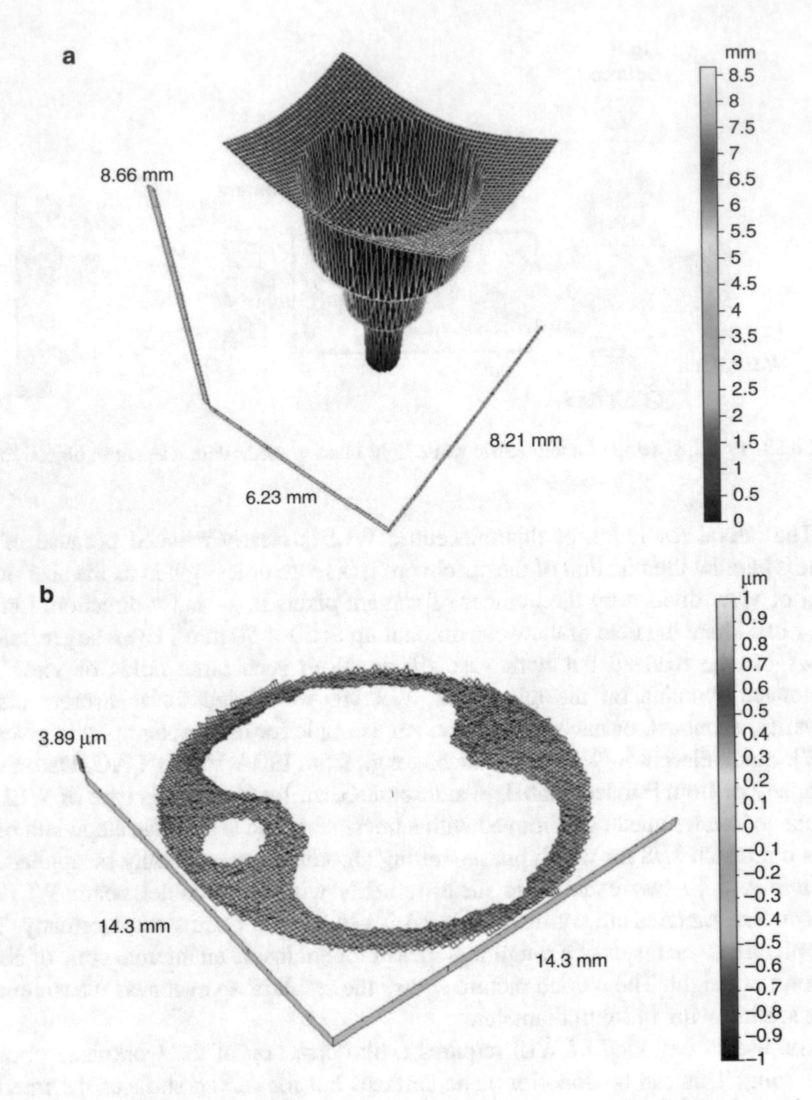

Fig. 5.40 Exemplaric measurements with a parallel beam WLI: (**a**) roughness in an internal cone, (**b**) eveness of a sealing. (Pictures courtesy of ISRA VISION AG, Darmstadt, Germany)

The task is now to determine the maxima of both envelopes. This task gets rendered more difficult if the thickness of the transparent film is as low as the correlograms are not well separated and will influence each other. A robust but not fast evaluation method can be to remove first the dc-component, then to square the

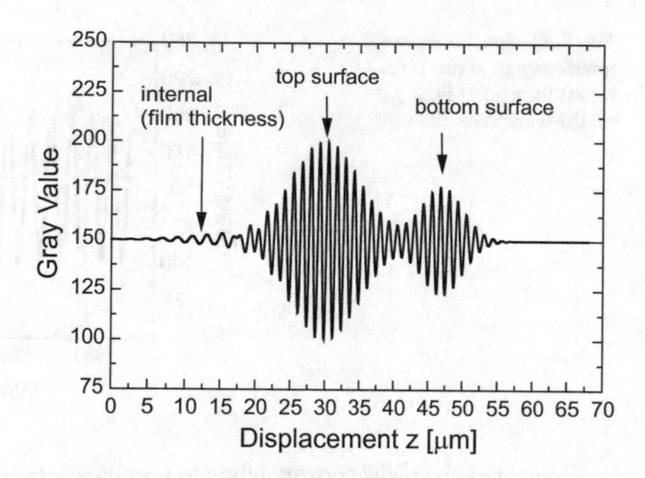

Fig. 5.41 Intensity correlogram of a substrate surface coated with a thin transparent film

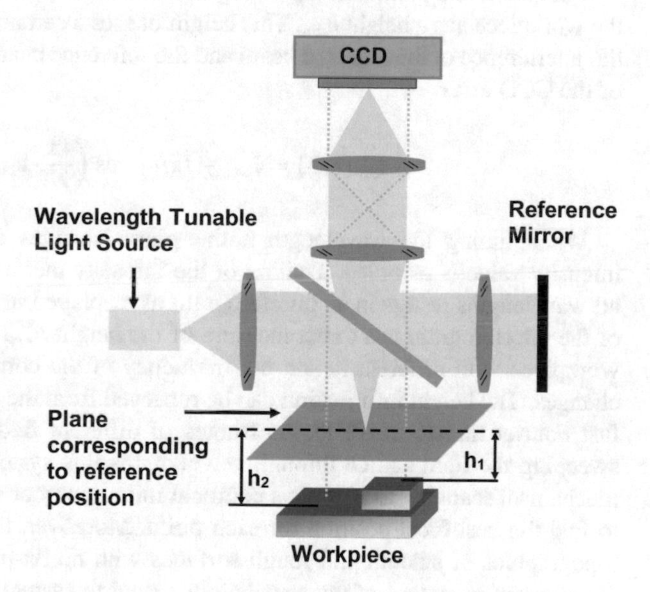

Fig. 5.42 Wavelength scanning interferometer

data, to retrieve the envelope from FFT with low pass filtering, and to use a fit of two Gaussian curves to the retrieved envelope.

5.5.4 Wavelength Scanning Interferometry

Wavelength Scanning Interferometry (WSI) takes advantage of spectral interference fringes for a wide range of wavelengths. These spectral fringes can be obtained when scanning the source wavelength or when dispersing white light with a grating. WSI is typically based on a Michelson interferometer and a tunable light source. The setup is shown in Fig. 5.42.

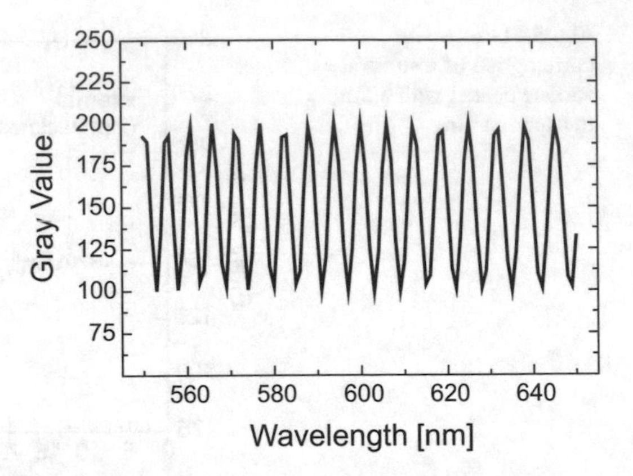

Fig. 5.43 Spectral intensity interferogram at one pixel (n, m) for a point $P(x_n, y_m)$ on the workpiece

Compared to a plane corresponding to a reference position the point $P(x_n, y_m)$ on the workpiece has a height h_{nm}. This height causes a certain phase in the intensity of the interference of the reflected beam and the reference beam recorded in pixel (n, m) of the CCD array.

$$\mathrm{I_{nm}}(\lambda, \mathrm{h}) = \mathrm{I_{DC}} + \mathrm{I_{AC}} \cdot \cos\left(\frac{4\pi}{\lambda} \cdot \mathrm{h_{nm}}\right) \tag{5.37}$$

When tuning the wavelength λ the phase changes accordingly and another intensity value is recorded. Trailing of the intensity measured at a camera pixel for all wavelengths results in an interferogram as displayed in Fig. 5.43. The frequency of the interferogram is a direct measure of the height h_{nm}. For another point on the workpiece with different height the frequency of the corresponding interferogram changes. The height information can be retrieved from the spectral interferogram by fast Fourier transform (FFT). As fringes of different frequencies are observed by sweeping the light source through wavelengths this system does not require axial mechanical scanning as in typical confocal microscopy or white light interferometry to find the best focal position for each point. Moreover, this measurement delivers topographies of smooth and rough surfaces with no 2π-phase ambiguity problem. One further advantage of the wavelength scanning system is that the contrast of the fringes remains good even for dispersive media.

The tuning range of the wavelength determines the resolution of the measurement and the tuning step of the wavelength determines the minimum and maximum measurable depth. In the beginning large expensive dye lasers and Ti:Sapphire lasers and more convenient tunable solid state lasers were used. Later, broadband sources like a superluminescent diode in combination with wavelength-tuning devices became more convenient and stable illuminating systems. With this technique submicron resolution is possible with a height range of a few millimeters. The height range is limited by the used finite wavelength range. The minimum measurable height is given by the wavelength range as

$$h_{min} = \frac{1}{2 \cdot \left(\frac{1}{\lambda_{min}} - \frac{1}{\lambda_{max}}\right)}. \tag{5.38}$$

The maximum measurable height is determined by the wavelength tuning step and the number *N* of scan points and follows from the Nyquist-Shannon theorem to

$$h_{max} = (N - 1) \cdot h_{min}. \tag{5.39}$$

In WSI using a Michelson interferometer there exists a height ambiguity problem. The reason is that the evaluation only considers the magnitude of the distance h_{nm} but not its direction. In the Michelson interferometer object height can be, in relative terms, on both sides of the reference mirror. In these cases special phase unwrapping procedures must be applied [108–113]. If the object is shifted above or below the focus plane no unwrapping procedure would be necessary because the object would be clearly on one side of focus or the other. Shifting the object away from focus increases the frequency of fringes at each point. This shifting may limit the maximum measurable height range. In a Fizeau setup multiple reflections between the reference and the object must be accounted for in the signal analysis. Moreover, the resultant signal is not sinusoidal as in a Michelson setup. The fundamental period of the signal is the same. In the Fizeau setup the object is on one side of the reference mirror and no ambiguity in height derivation exists.

Main applications for in-process surface metrology include:

- Large (> λ) discontinuous step heights measurement,
- V-groove measurement,
- Topography measurement on multilayer, thin or thick films,
- Thickness measurement of thin and thick films,
- Defect detection and identification of ALD coatings,
- Measurement of MEMS/NEMS systems, and
- Measurement of optics.

5.5.5 Multi-Wavelength Interferometry

Multi-Wavelength Interferometry (MWLI) provides a way to expand the capabilities of single wavelength interferometry with a synthesized longer wavelength. The single synthethic wavelength Λ is generated by using close-lying short wavelengths simultaneously. Obtained interferograms correspond to interferograms where a single wavelength, namely the synthesized or beat wavelength Λ, has been used. This technique enormously increases the range of decidedness or unambiguity for the distance determination compared to single wavelength interferometry, while the high resolution of up to 0.1 nm is maintained. The range of decidedness can amount up to 100 mm. While for single wavelength interferometry the interval of unambiguity

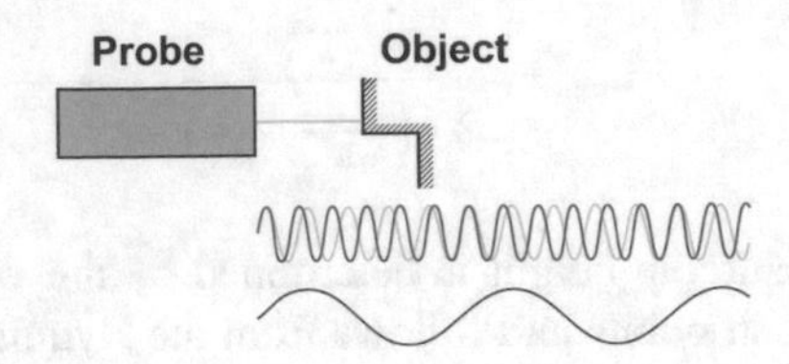

Fig. 5.44 Principle of increasing the range of unambiguity by using multiple proximate waves (blue and cyan wave) resulting in a beat with strongly enlarged beat wavelength (red wave)

is $\lambda/2$ it increases to $\Lambda/2$ for multiple proximate wavelengths. For example, for two wavelengths λ_1 and λ_2 with a small $\Delta\lambda = |\lambda_2 - \lambda_1|$ the beat wavelength is

$$\Lambda = \frac{\lambda_1 \cdot \lambda_2}{\Delta\lambda} \tag{5.40}$$

It is much longer than the original pair of wavelengths and becomes the longer the smaller the difference $\Delta\lambda$ is. Due to the large range of unambiguity also a large working distance can be established in MWLI.

In Fig. 5.44 the principle of increasing the range of unambiguity is sketched for two proximate waves (blue and cyan) and the resulting beat (red).

Interferometric techniques that employ two or more wavelengths have been described by several authors [114–118]. Petter [119] extended this technique to long wavelengths in the near infrared region and developed a distance sensor which is utilized in the system of AMETEK Germany GmbH – BU Taylor Hobson, Dept. Luphos, Weiterstadt, Germany. By default three wavelengths in the range between 1530 and 1610 nm are used.

The sensor head is shown in Fig. 5.45a. The used laser have a wavelength stability of several picometer. This high stability is necessary to achieve in a high accuracy and resolution. The accuracy of this head amounts to ±2 nm and a range of unambiguity of 1250 μm. The measuring head gets excited to oscillations of 2 kHz which in turn modulates the laser light accordingly. This modulation is used to evaluate the light reflected by the workpiece. Problems may arise with coatings that are transparent in the used wavelength range if their thickness is as small as light reflected at the rear side can enter the detector. For polished surfaces it is recommended to have the optical axis perpendicular to the surface within a tolerance angle of ±5°.

To achieve in the high resolution of ±2 nm in this interferometric method the sensor must be operated in a thermally stable and preferably vibration-free environment. Moreover, the absolute position of the sensor should be known as good as possible. Such an environment is given by a thermally stable housing of Invar steel equipped with two MWLI sensors that control the distances in vertical and horizontal direction and with the sample holder mounted on a granite base. Such a measuring system is schematically shown in Fig. 5.45b.

An application of the MWLI is the investigation of the surface of aspheric lenses. One example is shown in Fig. 5.46 for an aspheric lens with a diameter of 160 mm. The first picture shows the MWLI sensor in action followed by pictures of the

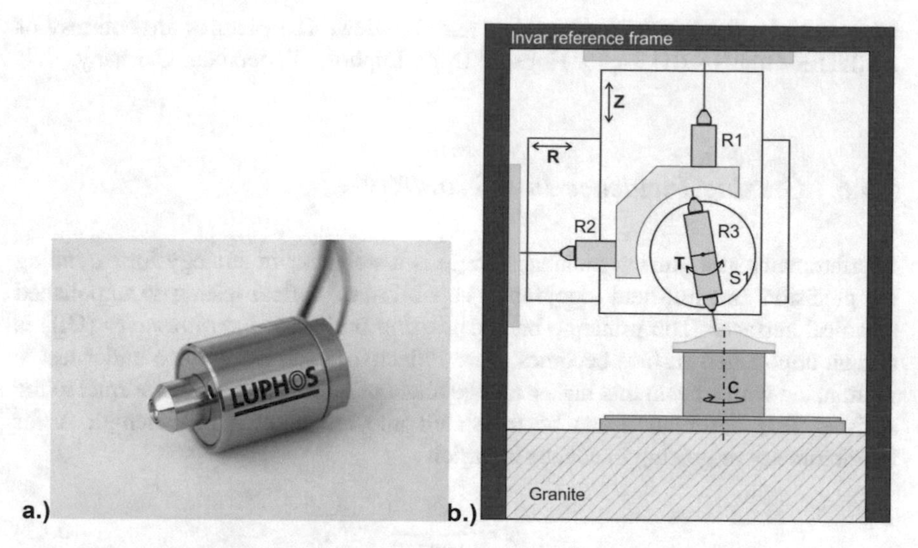

Fig. 5.45 (**a**) MWLI sensor, (**b**) MWLI measuring system of AMETEK GmbH – BU Taylor Hobson, Dept. Luphos, Weiterstadt, Germany

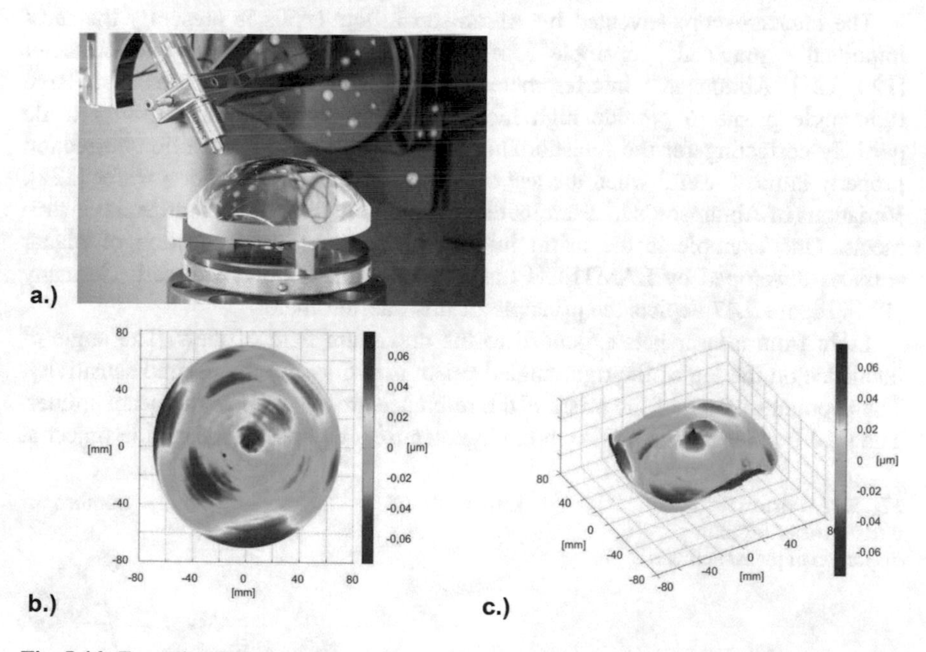

Fig. 5.46 Example of measurements with a MWLI sensor: surface topography of an aspheric lens with a diameter of 160 mm. (**a**) Photograph of the sensor in action, (**b**) Measured topography scan top view, (**c**) Measured topography scan 3D view. (Courtesy of AMETEK GmbH – BU Taylor Hobson, Dept. Luphos, Weiterstadt, Germany)

measured topography scan as top view and 3D view. The pictures are courtesy of AMETEK GmbH - BU Taylor Hobson, Dept. Luphos, Weiterstadt, Germany.

5.5.6 Grazing Incidence Interferometry

Interferometry at a grazing-incidence angle is a well-known strategy for extending the precision and full-field topography capability of optical testing to unpolished technical surfaces. The principle behind *grazing incidence interferometry* (GII) is that an unpolished surface becomes more reflective when the surface under test is illuminated with a beam that makes a large angle of incidence with the normal to the surface. This observation can be translated into an effective wavelength Λ for interferometry at grazing incidence in which

$$\Lambda = \frac{\lambda}{\cos(\alpha)} \tag{5.41}$$

where λ is the source wavelength and α is the angle of incidence.

The interferoscope invented by Abramson in the 1960s is presently the most important practical example of a grazing-incidence interferometer [120, 121]. Abramson's interferometer uses the refractive properties of a large right-angle prism to provide high incident angles on the surface (>80°) while partially correcting for the foreshortening of the image. This distortion correction property is most useful when the test object is circular, e.g. a silicon wafer [122]. Variations of Abramson's interferometer are common in modern commercial instruments. One example is the prism interferometer for form inspection of planar surfaces developed by LAMTECH Lasermesstechnik GmbH, Stuttgart, Germany [123]. Figure 5.47 depicts the principle of this interferometer.

Light from a laser gets expanded to the maximum field of view. The angle of incidence on the leg of the right-angled prism is defined by the desired sensitivity. The hypotenuse area of the prism is the reference area as well as the beam splitter. The light that leaves the prisma at the hypotenuse area hits the measuring object at

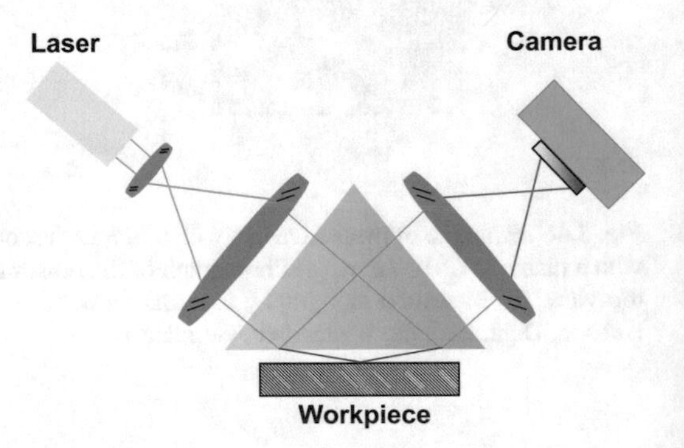

Fig. 5.47 Prism interferometer for form inspection of planar surfaces

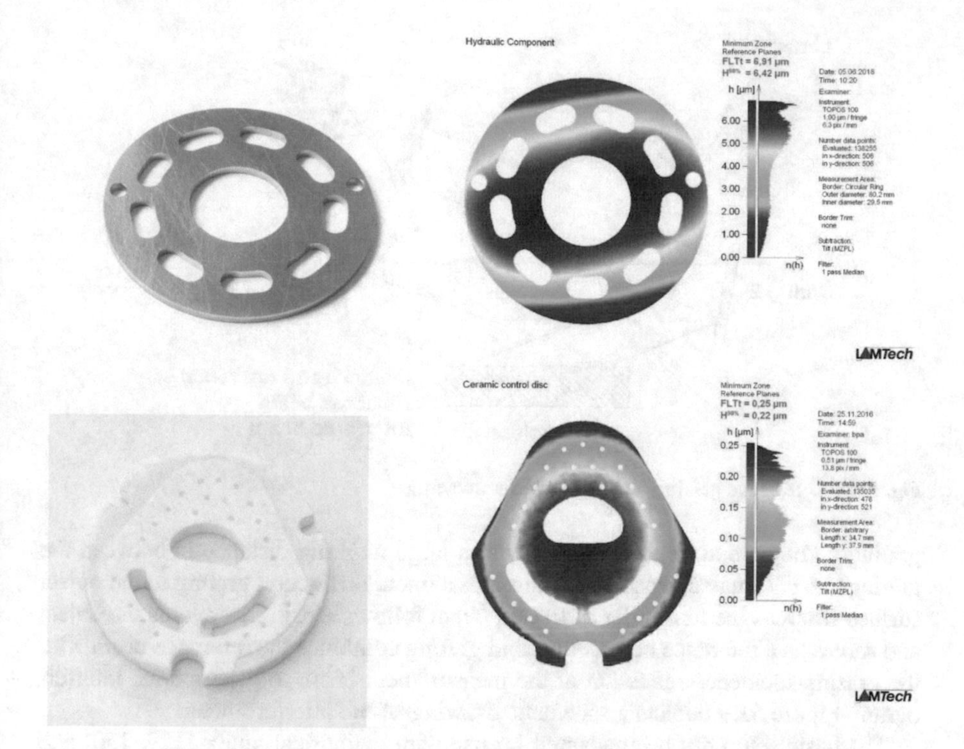

Fig. 5.48 Representative examples of measurement with a prism interferometer. Top: component of a hydraulic system, Bottom: single lever faucet ceramics. (Courtesy of LAMTECH Lasermesstechnik GmbH, Stuttgart, Germany)

grazing incidence and is reflected by the object. The reflected light interferes with the light reflected at the hypotenuse area. The interference pattern finally is projected on a camera. For consideration of the sample roughness the sensitivity can be selected from four calibrated settings (0.5, 1, 2, and 4 μm/fringe). The accuracy amounts to 0.1–0.4 μm peak-to-valley depending on the chosen sensitivity. Typical areas of application are the flatness measurement of lapped, flat-hones and fine-ground precision parts, valve plates, sealing and control discs, mechanical seals made of ceramics (Al_2O_3, SiC, Si_3N_4), metal (steel, aluminum, bronze), carbon or synthetics (duroplast, thermoplast, PMMA).

An important limitation to the design of a prism interferometer is that the test object must be placed a fraction of a millimeter away from the hypotenuse of the prism. This requirement follows from the desire to maximize the field of view while the aberrations associated with the steep refraction angle at the prism surface are minimized [124]. Exemplaric results obtained with the prism interferometer of LAMTECH Lasermesstechnik GmbH are shown in Fig. 5.48.

Another technique that has been developed in GII is the *diffractive grazing-incidence interferometer* [125–128]. This interferometer uses a pair of linear

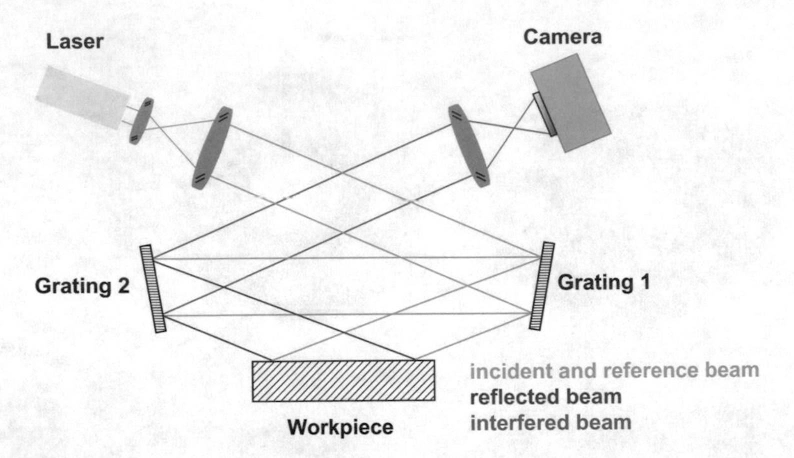

Fig. 5.49 Diffractive grazing-incidence interferometer

gratings. They are arranged in series with a large working distance in between the gratings. A collimated beam passes through a linear diffraction grating to the object surface whereas the first-order diffracted beam follows a path that avoids the object and serves as a reference beam. A second grating combines the reference beam with the grazing-incidence reflection of the measurement beam to generate the interferogram. Figure 5.49 depicts a schematic drawing of the interferometer.

The basic setup has been adapted for use with cylindrical optics [129, 130] and for phase-shifting interferometry [131]. This diffractive optic interferometer achieves good results over large areas with small-aperture optics and is particularly well suited to objects that are rather more long than wide. An important difficulty common to nearly all single-pass diffractive grazing-incidence geometries is that the reference and measurement beam wavefronts are inverted with respect to each other, leading to an undesired sensitivity to the wavefront uniformity and flatness. To circumvent these problems Peter de Groot [132] proposed a geometry for diffractive grazing-incidence that does not suffer from relative wavefront inversion between the measurement and the reference beams. This setup uses a fold mirror and needs only one linear transmission grating that serves as beam splitter and as beam combiner. A key element is a reference mirror (reference flat) which restores symmetry to the beam paths by equalizing the number of reflections and the optical path lengths.

The diffractive grazing-incidence interferometer can be used for single side inspection of surfaces up to currently about 200 mm at incident angle of 85°. One can achieve in an accuracy of 50 nm with repeatabilities in the order of 15 nm and a resolution of 5 nm. Roughness measurement is restricted to a maximum of $R_a \approx 1$ µm. The samples must have a reflectivity of at least 10% at incident angle 85°. This technique is used for example in the Tropel® Flat Master® products of Corning Inc., Corning, NY, U.S.A.

Extension of the basic setup to a configuration for measuring simultaneously the front and back surfaces of a flat object such as a silicon wafer was already done in 1972 by Birch [125]. Figure 5.50 depicts a schematic drawing of this configuration.

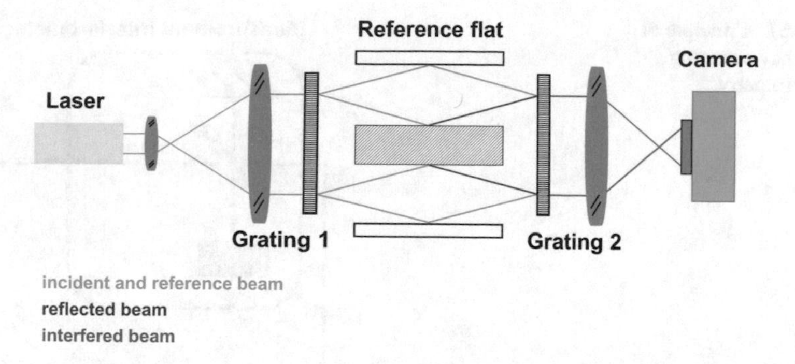

Fig. 5.50 Dual-sided diffractive grazing-incidence interferometer

5.5.7 *Frequency Scanning Interferometry*

Conventional single-wavelength interferometers offer excellent height resolution over a continuous smooth surface, however they cannot measure rough surfaces or the relative distance between discontinuous regions. *Frequency Scanning Interferometry* (FSI) [133–138] is a technique which allows to measure absolute distances to interferometric accuracy also on rough surfaces.

The principle of FSI is explained in the following and is depicted in Fig. 5.51. For each distance to be measured a measurement Michelson interferometer MM is built. The optical path difference L_{MM} of any MM is compared with the optical path difference L_{RM} of a stabilized reference Michelson interferometer RM by illuminating both interferometers with a tunable laser and tuning the frequency ν over a frequency interval $\Delta\nu$. Then, the phase change for RM is

$$\Delta\phi_{RM} = \frac{2\pi}{c} \cdot L_{RM} \cdot \Delta\nu \tag{5.42}$$

and similarly for MM

$$\Delta\phi_{MM} = \frac{2\pi}{c} \cdot L_{MM} \cdot \Delta\nu \tag{5.43}$$

The quotient of the measured phase changes yields the quotient of the optical path differences

$$\frac{\Delta\phi_{RM}}{\Delta\phi_{MM}} = \frac{L_{RM}}{L_{MM}} \tag{5.44}$$

This procedure becomes inaccurate due to drifts in both the reference and the measurement interferometer. Drifts are the main source of error in this method. Taking them into account Eq. (5.44) becomes

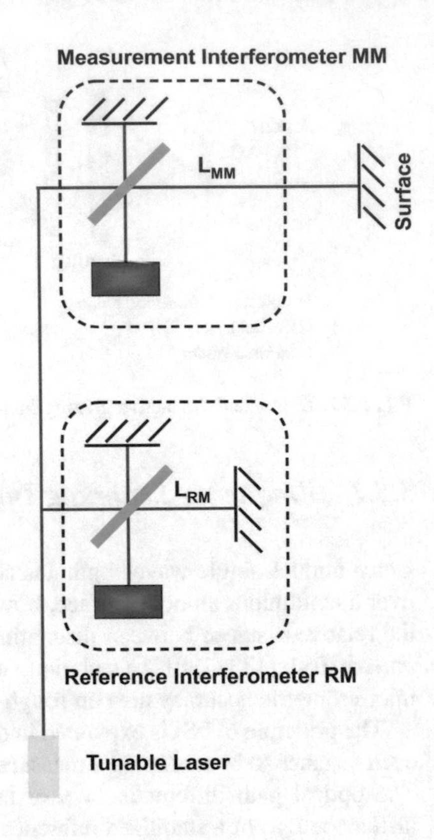

Fig. 5.51 Principle of frequency scanning interferometry

$$\frac{\Delta\phi_{RM}}{\Delta\phi_{MM}} = \frac{L_{RM}}{L_{MM}} \cdot \left(1 + \left(\frac{\Delta L_{RM}}{L_{RM}} - \frac{\Delta L_{MM}}{L_{MM}}\right) \cdot \frac{\nu_0}{\Delta\nu}\right) \tag{5.45}$$

where ν_0 is the center frequency of the laser in the interval $\Delta\nu$. Drifts as well as the influence of vibrations can be corrected using a second laser that illuminates the interferometers from opposite direction.

Early FSI systems used conventional bulk airspaced optics but recent research has evolved into the use of fiber-fed interferometers where the laser light is launched into single mode optical fibers and the output end of the fiber forms a miniature interferometer with the reference beam formed from the internal reflection at the fiber-air interface.

The big advantage of the method is that multiple distances can be measured simultaneously. The height of each pixel (= place of measurement) is obtained independent of its neighboring pixels using fast Fourier transform algorithms to determine the modulation frequency. No phase unwrapping algorithms must be used. As a result, it is possible to measure rough and diffuse surfaces in which there can be very large differences in height from pixel to pixel. This makes the method well-suited for measurements of surface flatness as well as the parallelism and height separation between surfaces. With two FSI sensors, data may be collected simultaneously on both sides of an object so that the total thickness also can be measured. The FSI technique allows for measurements of fields of view up to 300 mm with a height resolution of 1 nm. The lateral resolution (the pixel size) is restricted and is in the order of 40 μm for small fields of view and 150 μm for large fields of view. On the other hand, the measurement range for the height amounts to 50–300 mm. For flatness measurements accuracies of 60 nm and and repeatabilities of 20 nm are achievable, for parallelism 100 nm and 25 nm and for depth/height 250 nm and 100 nm. This technique is used for example in the Tropel® Flat Master® MSP products of Corning Inc., Corning, NY, U.S.A.

5.5.8 Digital Holographic Microscopy

The superposition of an object wave with a reference wave and storing the resulting interferogram in a photosensitive medium allows to have a complete complex image of the object, i.e. the complete information on phase and intensity at each point of the object. To retrieve this information one has to illuminate again this image with a reference beam. Dennis Gabor first demonstrated this principle of *holography* and Gabor invented a method to create images with complex fields as a means to transmit information over parallel channels [139–142]. Although Gabor originally proposed holography as a method for aberration correction of electron micrographs, optical holography took on a life of its own, with applications spanning from entertainment to secure encoding and data storage. In 1971 Gabor received the Nobel Prize in physics for this invention.

In *digital holography* a hologram is recorded by an electronic device and reconstructed digitally by simulating the reference beam. The idea was already published more than 30 years ago, for example by Goodman and Laurence [143]. Digital holography is now a well-established technique, e.g. [144–147], and has evolved to a flexible quantitative technique for the characterization of surface structures [148, 149]. *Digital holographic microscopy* (DHM) utilizes this interferometric technique for real-time imaging of the entire complex optical wavefront reflected by or transmitted through a sample with a vertical resolution of 1 nm in a single acquisition without a mechanical or spectral scan. This is a big advantage compared to vertical scanning methods (confocal microscope, focal depth variation, white light interferometry) with respect to vibrations since the time for acquisition of the complete image is in the order of milliseconds. So, a high measuring speed is achieved. The strength of the DHM lies on the use of the so-called off-axis configuration [150, 151] where the reference beam and the object beam enclose a small angle θ. Then, only one recorded hologram is needed while in on-axis configuration ($\theta = 0°$) multiple holograms must be recorded. The configuration of off-axis DHM and commercial DHM instruments are illustrated in Fig. 5.52 for reflection DHM (left) and transmission DHM (right). It is either a Mach-Zehnder configuration in reflection and transmission or a Michelson configuration in reflection.

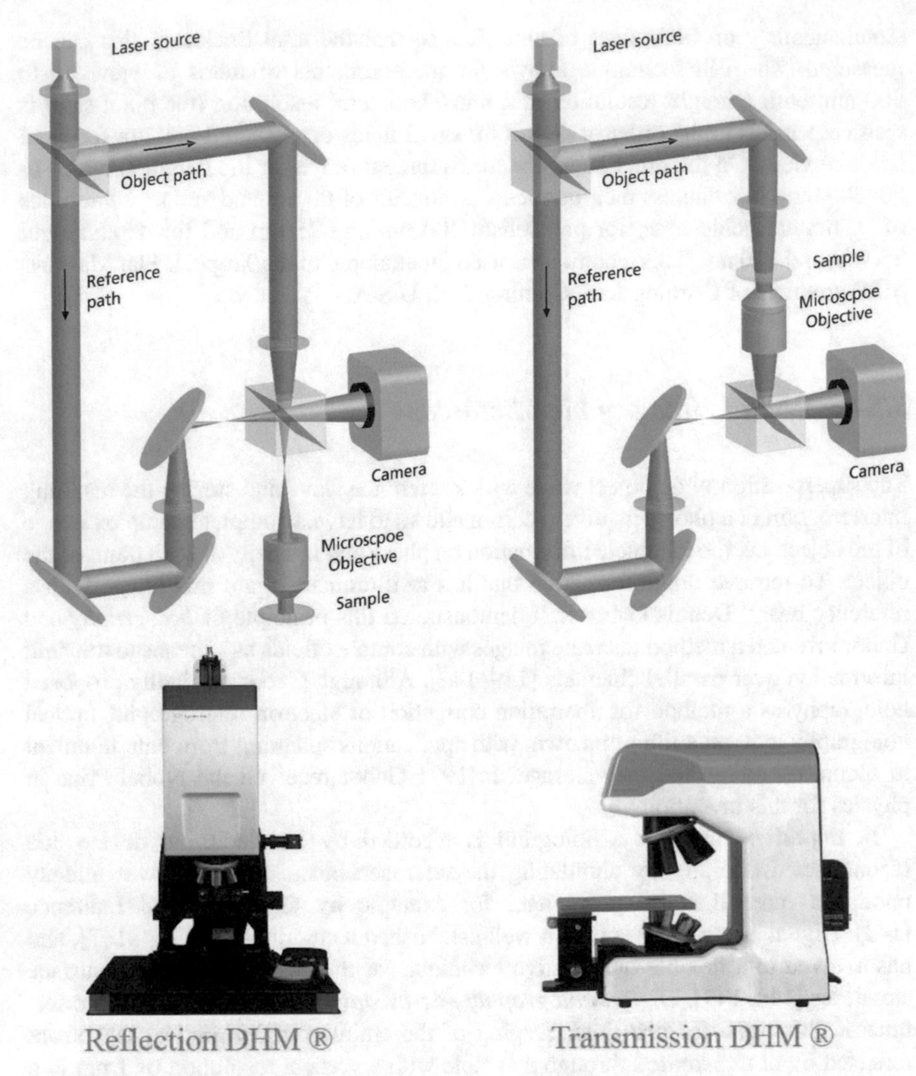

Fig. 5.52 Configuration of reflection DHM (left) and transmission DHM (right) and the corresponding commercial DHM instruments (bottom). (Courtesy of LyncéeTec SA, Lausanne, Switzerland)

The resulting interferogram from the beams in the object path and the reference path gets imaged on a CCD camera as digital hologram. To retrieve the 3D information on the spatial amplitude and phase distribution the digital hologram must be numerically reconstructed. There are a number of different methods developed for numerical calculation of the diffraction field. Common approaches in digital holography have been the Fourier transform, the Fresnel transform, and Huygens convolution methods as well as more recently the angular spectrum method [152, 153].

In principle, the reconstruction is done in three steps. The first step is to calculate the object wave in the plane of the hologram by solving the equation

$$I(x, y) = |R(x, y)|^2 + |O(x, y)|^2 + R^*(x, y) \cdot O(x, y) + R(x, y) \cdot O^*(x, y) \quad (5.46)$$

R(*x*, *y*) is the reference wave, *O*(*x*, *y*) is the object wave and the asterisk means the conjugated complex. In the second step the object wave gets reconstructed by numerical evaluation of the Fresnel-Kirchhoff diffraction integral in the plane of the measuring object. In off-axis digital holography the adjustment of the angle θ allows the separation of the different orders of diffraction in the frequency domain. So, the frequencies of the zero-order term ($|R(x,y)|^2 + |O(x,y)|^2$) and the twin-image term ($R(x,y) \cdot O^*(x,y)$) can easily be filtered to provide the real image [154, 155]. In the last third step a 3D phase map is generated on the computer. It contains the 3D information on the measuring object in intensity images and phase images.

In reflection mode the phase image reveals directly the surface topography with sub-nanometer vertical resolution while in transmission mode the phase image reveals the phase shift induced by the transparent specimen. The lateral resolution is determined by the field of view and the pixel number of the camera in a lensless setup or the choice of the microscope objective.

Problems arise with multiple layers in transparent samples and with scattering samples. This is one reason why DHM works well on smooth surfaces but cannot measure rough surfaces. Further problems arise with steep slopes and sharp edges. A surface with low arithmetic mean roughness R_a (R_a hundreds of nm) with local steep slopes and sharp edges will be more difficult to measure than a $R_a = 1$ µm surface with sinusoidal curvature. The 2π phase ambiguity in interference with coherent light restricts the measurement of large sample heights to the half wavelength in reflection and the wavelength in transmission. This can be improved using longer wavelengths but at the costs of the lateral resolution. When the object slopes are sufficiently small to have a continuous signal, also phase unwrapping procedures can overcome this ambiguity and enable the reconstruction of the topography. However, for abrupt steps greater than half of the wavelength (reflection), the phase unwrapping procedures can fail. Additionally, DHM cannot completely avoid parasitic interference effects and statistical noise such as shot noise.

The height measurement range can be extended using a dual wavelength acquisition with two wavelengths λ_1 and λ_2 with a small $\Delta\lambda = |\lambda_2 - \lambda_1|$. Then, the hologram corresponds to a hologram obtained with the beat wavelength

$$\Lambda = \frac{\lambda_1 \cdot \lambda_2}{\Delta\lambda} \quad (5.47)$$

This synthetic wavelength is much longer than the original pair of wavelengths.

In Fig. 5.53 two exemplaric measurements with the Reflection DHM® of LyncéeTec SA are depicted. The pictures are courtesy of LyncéeTec SA, Lausanne, Switzerland. The first picture in Fig. 5.53a demonstrates the high resolution

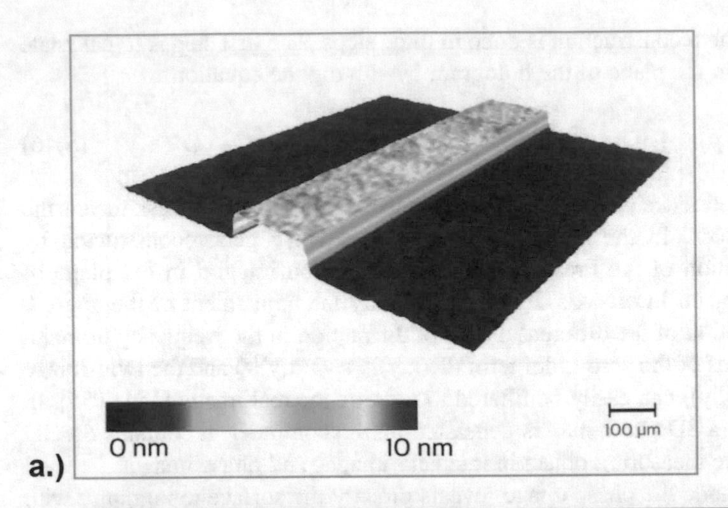

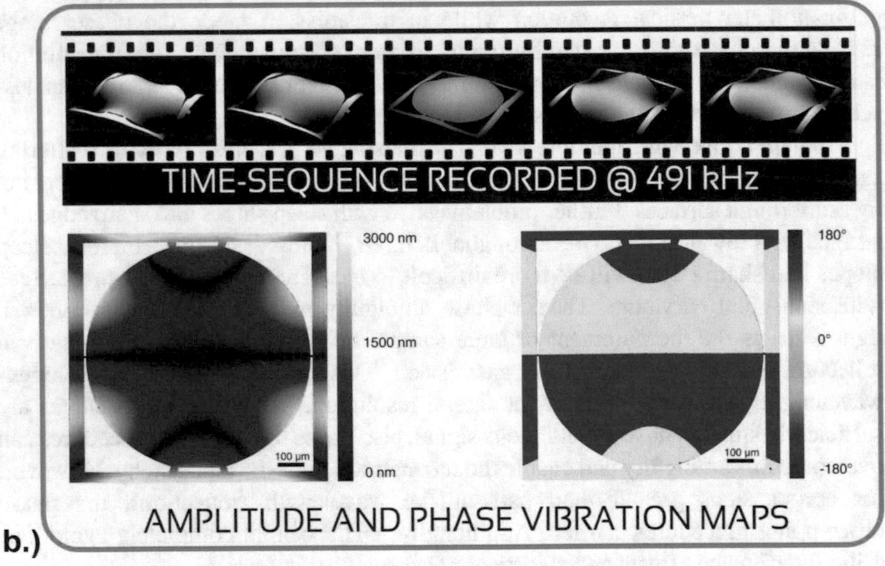

Fig. 5.53 Examples of DHM measurements with the Reflection DHM® of LyncéeTec SA. Courtesy of LyncéeTec SA, Lausanne, Switzerland

capability of digital holography showing the topography of a certified 9 nm step. The second picture in Fig. 5.53b demonstrates the capabilities of digital holography for time-resolved investigations. The picture shows a time-sequence of 3D topographies recorded at 491 kHz on a MEMS micro-mirror. The colored pictures are the amplitude and phase map of the picture in the middle of the sequence.

5.5.9 Conoscopy

In classical holography an interference pattern is formed between an object beam and a reference beam using a coherent light source. The object beam and the reference beam propagate with the same velocity but follow different geometrical paths. In *shearing interferometry* a wavefront is duplicated and the duplicated wavefront is laterally displaced by a small amount. The interference pattern is obtained from superposition of the original and the displaced wavefront. The most significant feature of the shearing interferometers is that they do not require a reference wavefront. Lateral shear interferometers have been used extensively in testing of optical components.

In *conoscopy*, a special kind of lateral shearing interferometry already proposed by Sirat and Psaltis [156, 157] in the 1980s and rigorously derived by Sirat [158, 159] in the early 1990s, the separate coherent beams are replaced by the ordinary and extraordinary components of a single beam traversing a uni-axial crystal placed between two crossed polarizers.

A conoscopic sensor is a point sensor that uses concentric optics for focusing a laser beam on the object and to image the reflected light on a CCD camera. The reflected light passes the actual conoscopic module of polarizers and birefringent crystal in front of the CCD camera. The principle of a conoscopic point sensor is shown in Fig. 5.54.

The shift of the ordinary and extraordinary beam produces ring-like holograms on the detector even with incoherent light, with fringe periods that can be measured precisely to determine the exact distance to the point measured. If R_m is the radius of the m-th fringe the distance d between sensor and and the surface is given as

$$d = \sqrt{\left(m\frac{\lambda}{2}\right)^2 - R_m^2} \tag{5.48}$$

The setup is insensitive to the position of the key optical components. The measuring range in axial direction can be changed by changing the objective. For

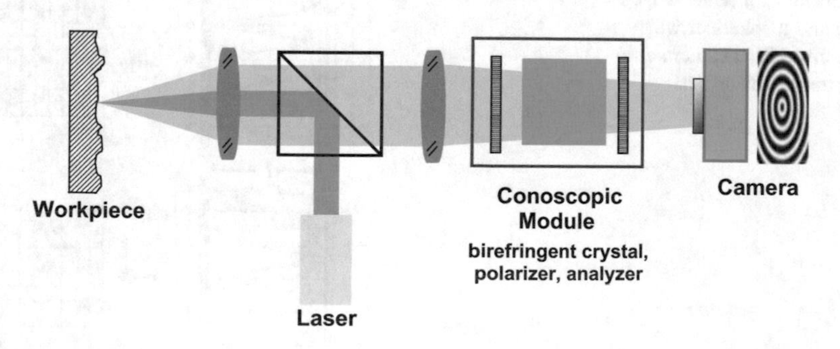

Fig. 5.54 Principle of a conoscopic sensor

example, for the conoscopic sensor of Optimet, Jerusalem, Israel, the measuring range is $z = 1.8$ mm, 8 mm, and 18 mm for focal length of the objective of $f = 25$ mm, 50 mm, and 75 mm. Vertical resolutions of better than 0.1 μm are obtainable.

5.6 Wave Front Sensing (Shack-Hartmann)

A *Shack-Hartmann sensor* detects the inclination of a wavefront in two directions x and y and from this primary data the wavefront can be reconstructed numerically. In the fundamental principle the Shack-Hartmann sensor can work without an external reference what makes it fast and insensitive to vibrations. Reference measurements can be taken to achieve in a high accuracy and can be subtracted to take into account for example static errors of the test instrument.

The incoming wavefront is divided into subparts by a microlens array. The light within each microlens is focused on a subarray of pixels of a camera chip. This is exemplarily illustrated in Fig. 5.55 for a planar and a spherical wavefront.

The Shack-Hartmann sensor so transforms phase information into a measurable intensity distribution. The lateral shifts of the focal length of each microlens are measured and evaluated. If σ_x and σ_y are the shifts in x- and y-direction, the gradients α_x and α_y are obtained from

$$\alpha_x = \frac{\sigma_x}{\sqrt{\sigma_x^2 + \sigma_y^2 + f^2}}; \quad \alpha_y = \frac{\sigma_y}{\sqrt{\sigma_x^2 + \sigma_y^2 + f^2}} \tag{5.49}$$

where f is the focal length of the lenses in the lens array. The wavefront gets reconstructed from this field of gradients. For evaluation the reconstructed wavefront is decomposed into a linear combination of Zernike polynomials (see again Sect. 5.5.1.2 for the Zernike polynomials).

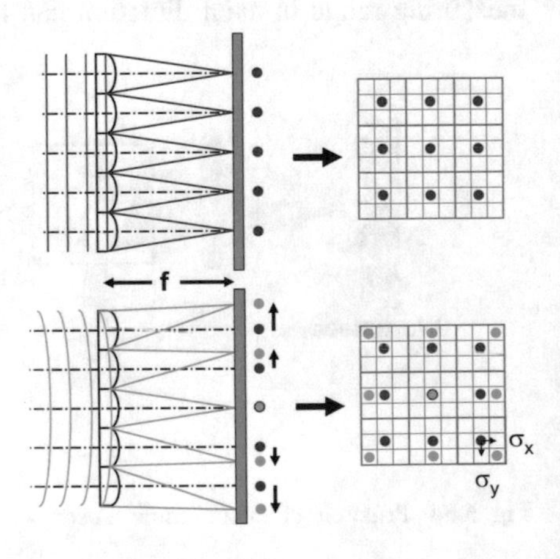

Fig. 5.55 Projection of the wavefront of a plane wave (top) and a spherical wave (bottom) onto a camera chip by a microlens array

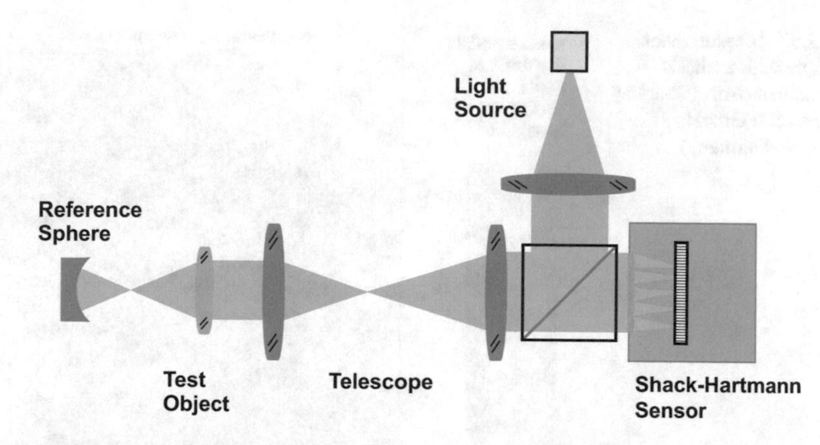

Fig. 5.56 Optical setup for quality control of lenses using a Shack-Hartmann sensor

For a planar wavefront each of the lenses generates a spot in the focal plane in the center of the pixel subarrays. These spots can be defined as reference positions on the camera. In case of a spherical wavefront the spots are in good approximation equidistant but their distance can be larger or smaller than in the case of a planar wavefront. In case of more general wavefront shapes the spot pattern will be less regular and can be distorted depending on the content of aberrations in the wavefront.

To separate the spots on the camera their diameter should not be larger than half the distance between two microlens centers. This condition also determines the maximum focal length of the microlenses. The larger the focal length is the larger is the sensitivity of the sensor. Typical Shack-Hartmann microlens arrays have between 10×10 and 100×100 microlenses with lens diameters between 0.2 and 2 mm. Typical repeatabilities and accuracies of Shack-Hartmann wavefront sensors are 1–2 nm r.m.s. and 4–5 nm r.m.s. respectively.

Nowadays, the Shack-Hartmann sensor is applied in astronomy (e.g., measurement and correction of air turbulence effects of the atmosphere in observatories), in ophthalmology (e.g., measurement of aberrations of the human eye), or in industry (e.g., measurement of the quality of microscope objectives, mobile phone objectives, ophthalmic lenses, etc.). In industrial applications the main fields of application are in direct measurement of laser beams or in testing optical elements in single pass, double pass or in reflected light measurement configurations. As an example Fig. 5.56 shows the schematic optical setup for quality control of lenses in transmission.

A corresponding measurement result is shown in Fig. 5.57. It shows the measurement of a mobile phone lens with an optical test instrument based on a Shack-Hartmann sensor as shown in Fig. 5.58. The tilt and defocus terms have been subtracted so the remaining asymmetrical wave aberrations can clearly be recognized. With a good calibration such instruments can achieve an accuracy of $\lambda/20$

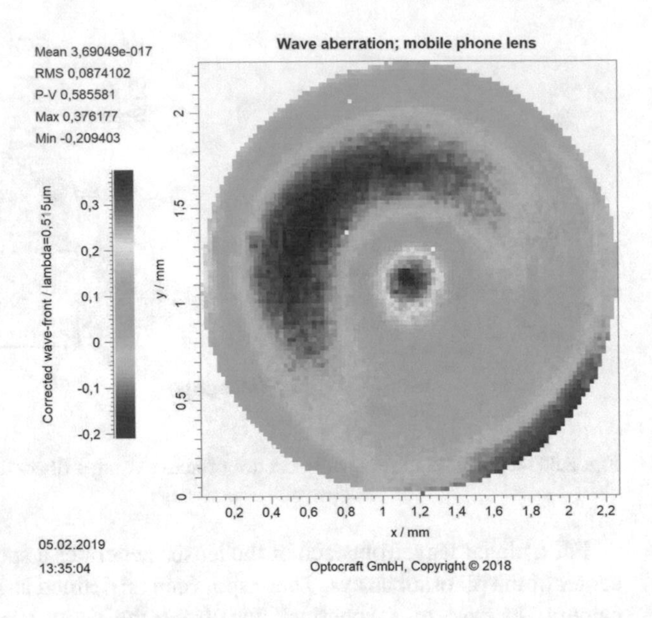

Fig. 5.57 Measurement example with a Shack Hartmann sensor. (Courtesy of Optocraft GmbH, Erlangen, Germany)

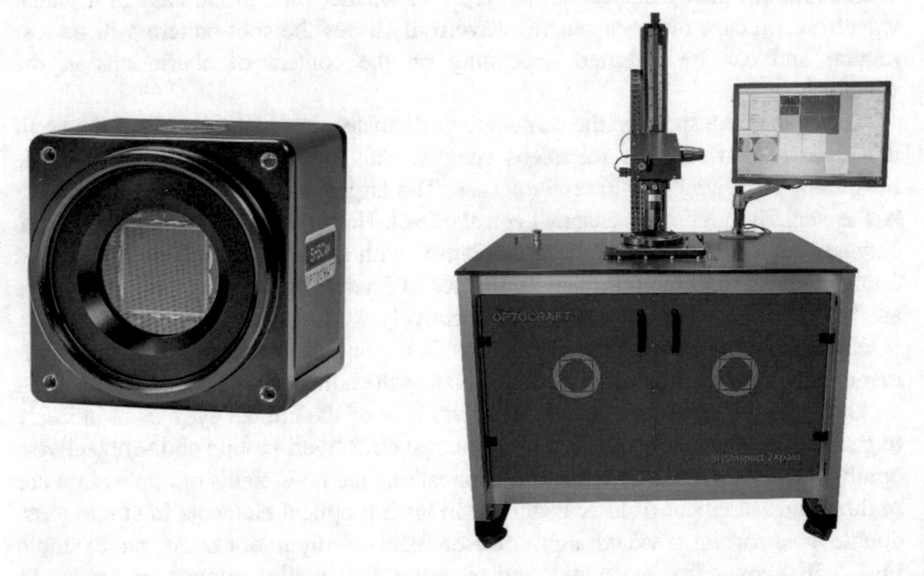

Fig. 5.58 Left: Shack-Hartmann wavefront sensor SHSLab. Right: SHSInspect 2Xpass for the measurement of lens systems as microscope lenses, mobile phone lenses, etc. (Courtesy of Optocraft GmbH, Erlangen, Germany)

peak-to-valley which is adequate for many industrial applications. The measurement and pictures in both Figs. are courtesy of Optocraft GmbH, Erlangen, Germany.

In the basic principle of the Shack-Hartmann sensor its accuracy and dynamic range are not independent of each other. If y_{min} is the smallest detectable shift the corresponding smallest tilt of the local part of the wavefront is

$$\theta_{\min} = \frac{y_{\min}}{f} \tag{5.50}$$

On the other hand the maximum detectable tilt is

$$\theta_{\max} = \frac{y_{\max}}{f} = \frac{d}{2f} \tag{5.51}$$

Here, f and d are the focal length and the diameter of a lens in the lens array. To increase the accuracy one must increase the focal length f according to Eq. (5.50). Then, the dynamic range automatically decreases according to Eq. (5.51). Therefore, accuracy and dynamic range must be adjusted accordingly to the application.

The Shack-Hartmann sensor was already developed in 1900 by J. Hartmann [160] as test method for disks with pinholes. In 1971 Shack and Platt built a fully functional sensor [161, 162]. Later it has been introduced in applications in optical shop testing and others [163].

Modifications of the Shack-Hartmann sensor are based upon the use of a Liquid Crystal Display (LCD) as switchable slit in front of one single lens with high numerical aperture instead of a lens array [164]. In a second setup with LCD the LCD operates as slit as well as diffractive optical element [165]. At each point of the LCD a diffractive structure in form of gray levels is displayed that acts as holographic lens. A third method [166] utilizes digital micro-mirror devices (DMD) where each mirror functions as adjustable reflecting slit. The selective switching of a single micro-mirror allows for a fast rasterizing of the wavefront.

5.7 Deflectometry

Deflectometry utilizes the deformation of a projected sinusoidal fringe pattern after reflection from a test specular surface to infer topographical information on the surface. So far, it is pretty similar to the fringe projection method in Sect. 5.3.3. On the other hand, fringe projection depends on diffuse reflection and cannot be simply used for the assessment of specular surfaces without additional effort. Deflectometry closes this gap in inspection and measurement technology. The principle of a deflectometric measurement is depicted in Fig. 5.59.

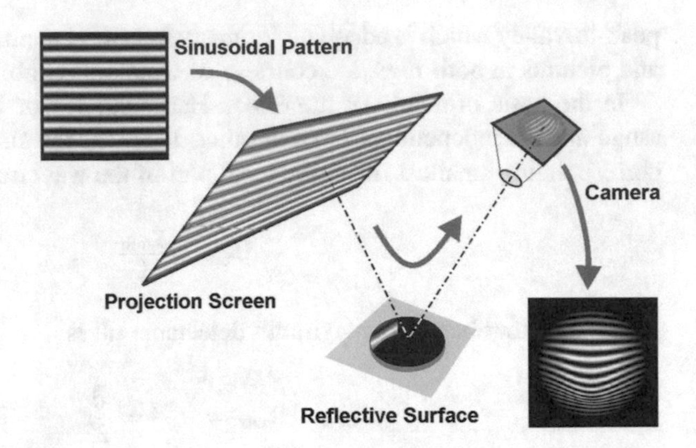

Fig. 5.59 Deflectometric inspection principle. (Courtesy of ISRA VISION AG, Darmstadt, Germany)

An illumination unit projects a sinusoidal fringe pattern on a diffusing screen. The reflected pattern on the surface of the measuring object gets detected with a camera. Any deviations from a flat surface give rise to a distortion of the observed fringes. Knowing all orientation parameters between camera and display it is possible to calculate the normal vector at each surface point and from this the gradient in the surface curvature from the measured phase distribution. The form of the surface follows from the gradient data by integration or differentiation [167–169]. Hence, a deflectometer does not measure local heights as other 3D sensors. Deflectometry is in general not unambiguous in the calculation of the normal vector to the surface. Any information on the local height is lost since only the gradient is of interest. This unambiguity is canceled out by using two cameras and phase-shifting deflectometry [170–173]. In phase-shifting deflectometry a series of phase-shifted fringe patterns are projected and for each pattern an image is recorded.

As the measurable area on the surface strongly depends on the position and shape of the object, the geometrical configuration of the measurement system must be adjusted to the object, i.e. camera focal length, CCD size, display size and orientation, etc. must be correctly selected and the deflectometer must be well calibrated. Then, the obtainable accuracy corresponds to a few nanometers in height difference. Cycle times for an area of 200 mm × 300 mm lie typically in the order of 10–30 s, with a measuring time of 5–15 s followed by a an evaluation with 5–15 s duration.

This incoherent and robust method is well-suited for the inspection of surfaces used in the manufacturing industry or for surfaces that just need to look good, e.g. car body components, to detect pretty fast form deviations as waves, dings, bumps, or scratches. Figure 5.60 depicts two typical measurements with a deflectometer. The first picture shows the measurement on an exterior car mirror. The second picture shows the result of the flatness measurement of a silicon wafer. The pictures are courtesy of ISRA VISION AG, Darmstadt, Germany.

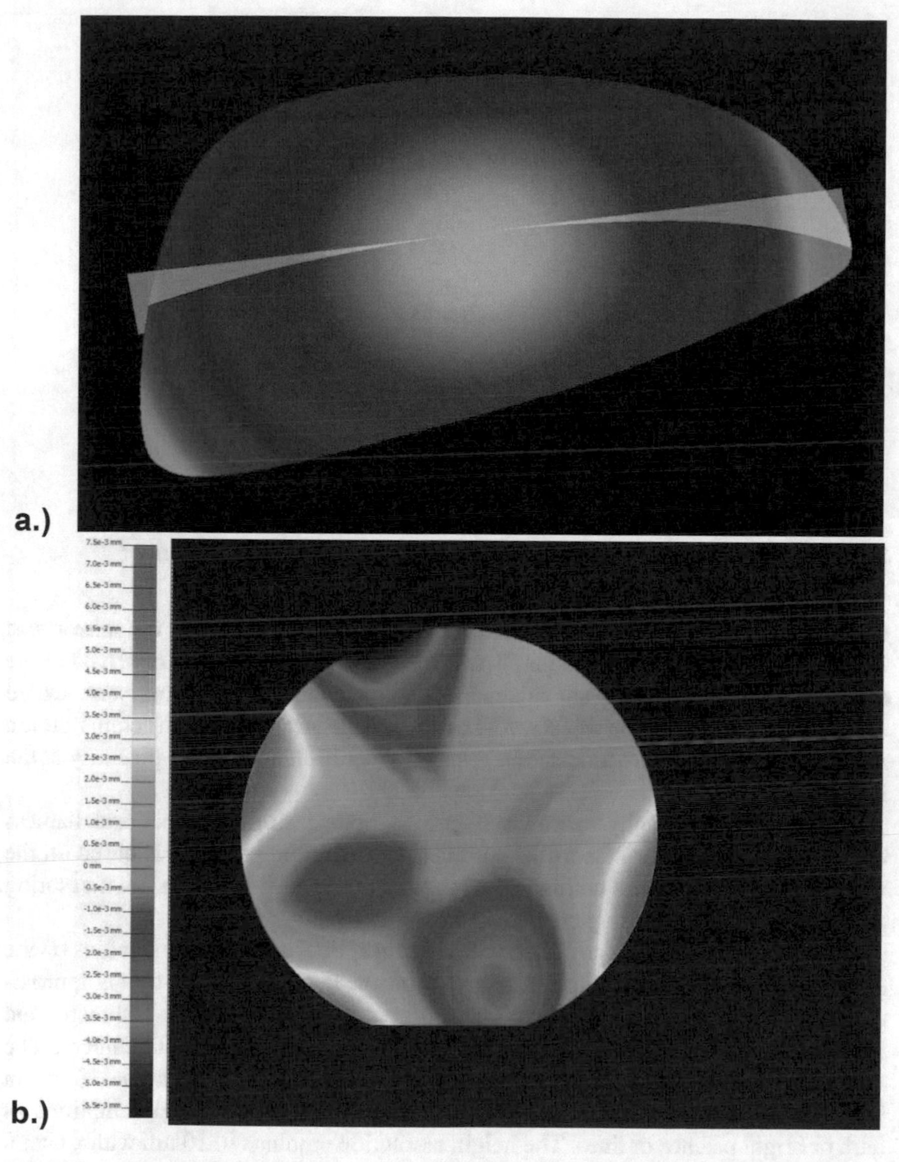

Fig. 5.60 Deflectometric measurements of (**a**) an exterior car mirror (with profile), (**b**) the flatness of a silicon wafer. (Courtesy of ISRA VISION AG, Darmstadt, Germany)

5.8 Makyoh Topography Sensor

The *Makyoh topography sensor* has its origin in the ancient China more than 2000 years ago. During the Han dynasty mirrors of bronze were manufactured with incorporated print characters and reliefs on the rear side. These structures

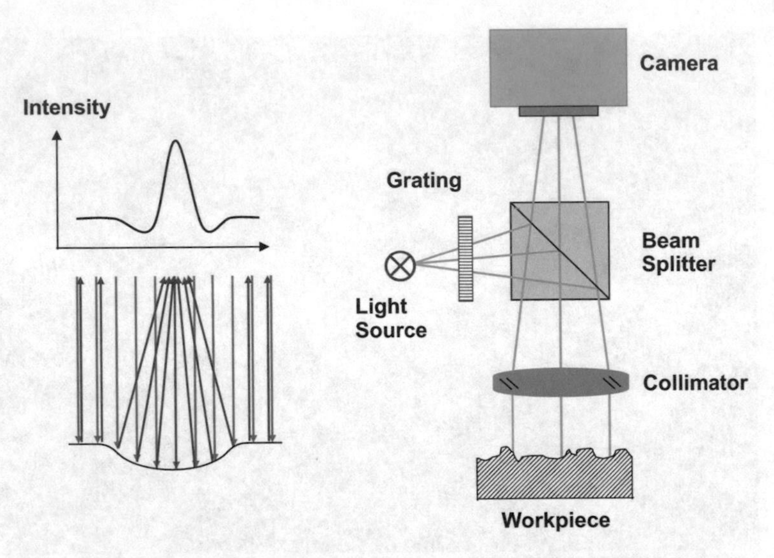

Fig. 5.61 Measuring principle (left) and schematic setup (right) of a Makyoh sensor

could not be seen on the highly reflective front side except of that the mirror was illuminated with (parallel) sunlight. Then, these structures could be observed at the opposite wall. Due to this unexplainable phenomenon these mirrors were called *magic mirror*, in Japanese *Ma-Kyoh*. Today, it is known that the structures on the rear side of the mirror are reflected in microscopic differences of the planarity at the front side.

A Makyoh topography sensor is hence an optical surface defect and flatness characterization tool for mirror-like surfaces. Its operation principle is based on the whole-field reflection of a collimated beam and defocused detection. Its measuring principle and principle setup are shown in Fig. 5.61.

The light of a LED gets projected with a beam splitter and a collimator lens on the surface of the object as collimated parallel beam. It gets reflected by an approximately flat, highly polished surface. The reflected beam traverses the collimator and the beam splitter and generates an image of the surface on a CCD camera. The recorded intensity distribution contains information about all deviations from an ideal flat surface. Even small height variations will show up strongly amplified as dark or bright patches or lines. The height resolution amounts to 10 nm with a lateral resolution of 50 μm. Surface areas of up to 50,000 mm^2 can be investigated. The accuracy is also determined by the used collimator optics and can amount to a few nanometers. Makyoh topography sensors have been used until now for a number of years as a sensitive tool for the inspection of mirror polished surfaces and in particular semiconductor wafer surfaces (e.g. [174, 175]).

One drawback of a Makyoh sensor is that the height profile cannot be retrieved unambiguously from the recorded Makyoh image [176]. For a quantitative height determination, the setup must be extended. One possibility is to project a well-known pattern on the surface and to analyze the deformation of this pattern after

Fig. 5.62 Makyoh topography image of a plane reference mirror (left) and a deformed Si wafer surface (right) after projection of a regular pattern on the surface

reflection. This method is quite similar to deflectometry. Another possibility is to project the diffraction pattern of a grating on the surface. The evaluation of the distorted wavefront of diffracted light is rather similar to that of Shack-Hartmann sensors. Figure 5.62 shows an example of the evaluation of a deformed wafer surface using pattern projection.

5.9 Surface Profiling Using Elastic Light Scattering

Elastic light scattering at surfaces is an area-integrating method usually carried out as power measurement with loss of information on the phase. Hence, to retrieve the complete topography of the surface from the scattered power is like to retrieve a clear image of a dragon from its traces. Nevertheless, certain surface characteristics like the r.m.s. roughness parameter R_q can be deduced from scattering as a characteristic of the area as a whole. Scatterometry is applied in surface metrology to detect surface contaminations, to characterize diffusely scattering workpieces or periodical structures (compact discs, structured photoresists, deposited structures), and to determine surface roughness parameters.

For surface roughness characterization it can be distinguished between conventional light scattering techniques and speckle techniques, both with further sub-techniques, as illustrated in Fig. 5.63.

Yet, the use of light scattering techniques for determination of r.m.s. roughness parameters is restricted. Vorburger et al. [177] identified several regimes that permit measurement of different surface parameters and functions using light scattering. They also established approximate limits for each regime. The regimes are illustrated in Fig. 5.64.

Please, also note again that the r.m.s. roughness (R_q or S_q) calculated from a model using scattered light measurements is not identical to the R_q or S_q value calculated from a profile or a topography image.

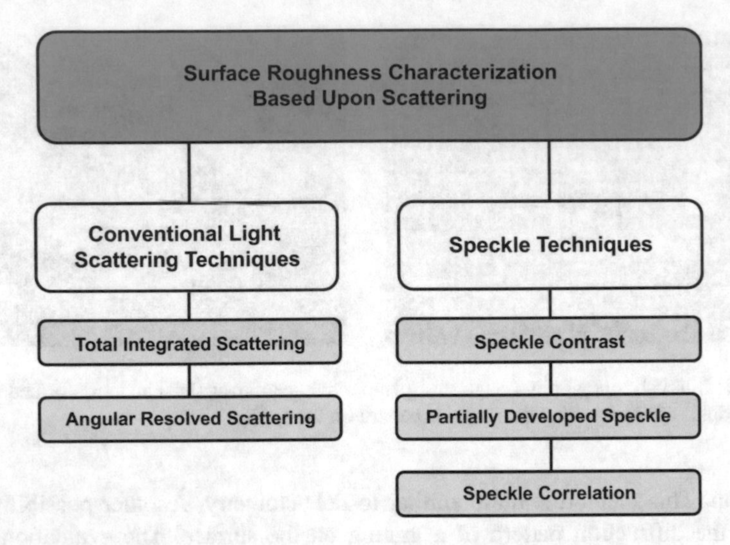

Fig. 5.63 Surface roughness characterization based upon scattering

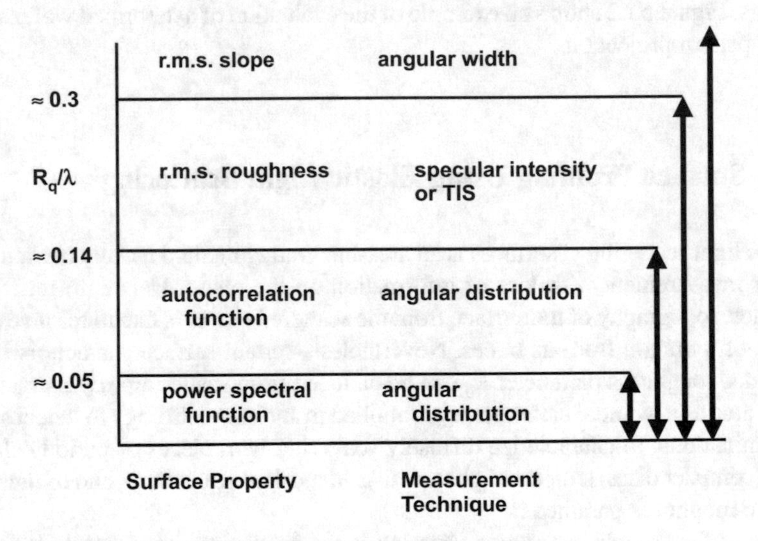

Fig. 5.64 Schematic diagram showing regimes of r.m.s. roughness R_q and measurable surface properties that can be derived from light scattering measurements

5.9.1 Total Integrated Scattering (TIS)

The determination of the surface roughness by *total integrated scattering* (TIS) is extensively described in the ASTM Standard F1048–87R99 or the SEMI MF 1048–1109 [178, 179]. Here, a brief description is given.

The TIS utilizes either a Coblentz sphere or an integrating sphere for collecting the light scattered by the surface into the upper hemisphere in a focus point where a

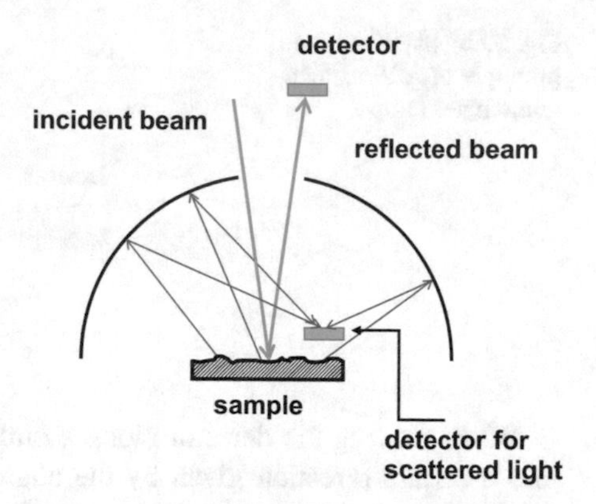

Fig. 5.65 Sketch of the TIS principle in reflection using a Coblentz sphere

photodetector measures the intensity. Figure 5.65 shows a sketch of the TIS principle with a Coblentz sphere. When using an integrating sphere scattering can also be measured in transmission.

The angle of incidence θ_{inc} is usually small ($\theta_{inc} < 10°$, mostly $\theta_{inc} = 5°$). The specular reflected light leaves the Coblentz sphere with the same small angle and gets detected with a detector outside the Coblentz sphere. This assures that only scattered light within almost the whole spherical half space gets detected by the sensor inside this half space.

For evaluation of the measured intensity again the functional relationship of Bennett and Porteus [180] between TIS and surface roughness

$$\mathrm{TIS} = \mathrm{R}_0 \cdot \left(1 - \exp\left(-\left(\frac{4\pi \mathrm{R}_q \cos\theta_{inc}}{\lambda}\right)^2\right)\right). \quad (5.52)$$

can be used. In this equation R_0 is the theoretical reflectance of the surface. TIS measurements are well-suited for determination of $R_q < 150$ nm.

In semiconductor industry TIS is an established method for surface defect detection. The surface is scanned with a laser beam and the light scattered in one or more finite angular regions is detected. This is used to detect residues, scratches, and particles and to determine haze.

5.9.2 Angular Resolved Scattering (ARS)

Angular resolved scattering (ARS) uses either a goniometric movement of a photodetector or a photodiode array to detect the light scattered into a certain angular range. The ASTM Standard E2387–05 [181] and the SEMI ME 1392–1109 [182] describe the procedures for this light scattering. Figure 5.66 shows the principle of ARS when using a goniometer setup.

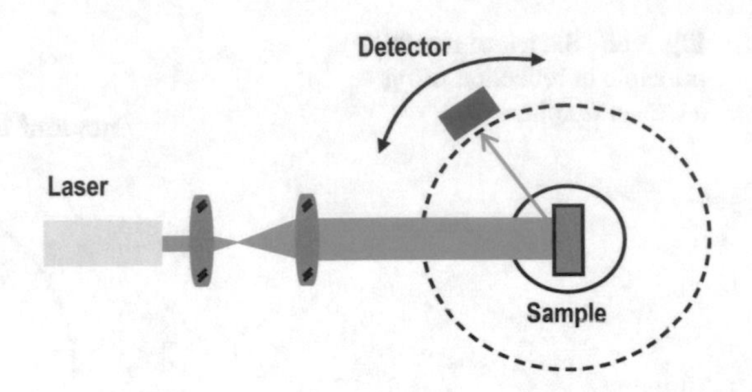

Fig. 5.66 Sketch of the principle of ARS using a goniometer setup

When moving the detector along a circle around the sample the light scattered into a certain direction given by the angle θ is measured. This yields the spatial distribution of the scattered light in the observation plane.

ARS is well-suited for the investigation on finished surfaces and on glossy and specular surfaces. It is inappropriate for matte surfaces, surfaces with a strong diffuse scattering, or coarse surfaces. The evaluation of the distribution of the scattered light allows to determine the quadratic average of profile gradients Δ_q (see Eq. (1.18)) which is related to the ratio of the r.m.s. roughness R_q and the correlation length Λ_k (see Eq. (1.21)) R_q/Λ_k. With ARS measurements $R_q < 150$ nm can be determined.

According to the ASTM Standard E2387–05 [181] the ARS is identical to the bidirectional reflection distribution function BRDF scaled by a factor of $\cos(\theta_{sca})$.

$$\mathrm{ARS} = \mathrm{BRDF} \cdot \cos(\theta_{sca}) \tag{5.53}$$

so that the power spectral distribution (PSD) can be calculated from the ARS measurement (see again Sect. 4.10 for the corresponding relations between PSD and BRDF). An example of ARS measurements and the calculated PSDs is given in Fig. 5.67 for a sample with low, medium, and high r.m.s. roughness parameter R_q. The roughness parameter is obtained from the integral of the PSD curve which increases with increasing roughness value. The form of the PSD curve depends on the actual distribution of roughness on the surface.

Although angular resolved scattering (ARS) is a well-known technique with many publications on it, it is rarely realized in commercially available measurement systems. A goniometric setup is too voluminous. Instead, the light scattered in a certain angular range along a line is captured and the recorded intensity distribution is evaluated for roughness parameters. This profile measurement is not just a special form of the area spectrum but is related to it in a more complicated way than might have been supposed. A rather useful model to relate the profile measurement with the area spectrum is known as K-correlation or ABC model [183].

In the system of OptoSurf GmbH, Ettlingen, Germany [184, 185], an objective with high numerical aperture captures the light scattered into only a small defined angular region with a diode line detector. From the recorded intensity distribution $H(\varphi_n)$ of scattered light at angles φ_n, $n = 1, 2, ..., N$, new surface characteristics (A_q, A_{sk}, A_{qm}, etc.) are retrieved. They are often pretty good related to the mode of operation, to manufacturing data, or to tactile roughness parameters R_a and R_z.

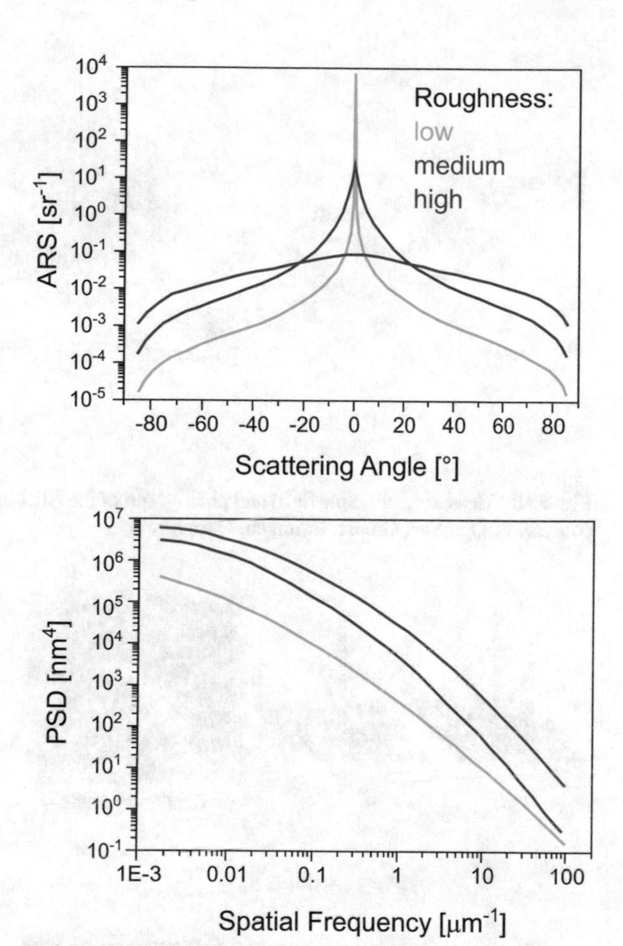

Fig. 5.67 ARS (top) and the corresponding PSD (bottom) for three differently rough surfaces with low, medium, and high roughness

Unfortunately, a direct determination of roughness parameters like R_a or R_z from the distribution of light is impossible. The measuring principle and a photograph of this ARS sensor are shown in Fig. 5.68. The pictures are courtesy of OptoSurf GmbH, Ettlingen, Germany.

The most important new characteristics are A_q and M. They are defined as

$$\mathrm{A_q} = \mathrm{k} \cdot \sum_{\mathrm{n}=1}^{\mathrm{N}} (\varphi_\mathrm{n} - \mathrm{M})^2 \cdot \mathrm{H}(\varphi_\mathrm{n}) \tag{5.54}$$

and

$$M = \sum_{n=1}^{N} \varphi_n \cdot H(\varphi_n) \tag{5.55}$$

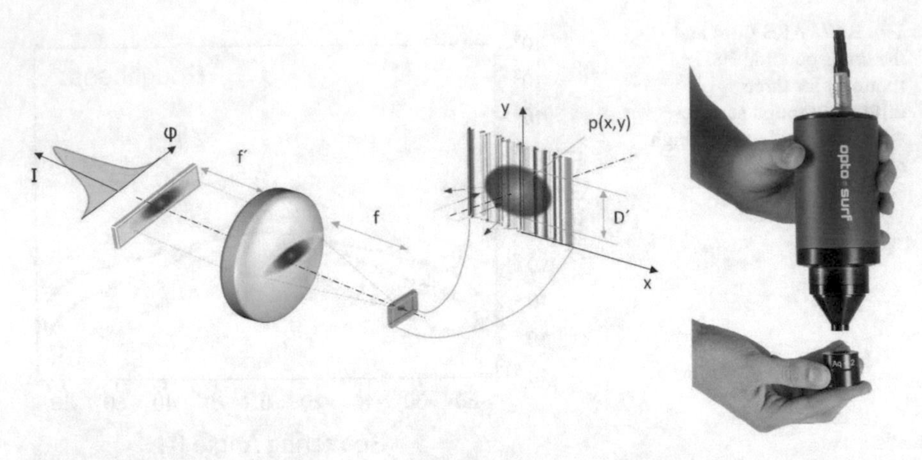

Fig. 5.68 Measuring principle (left) and photograph of the ARS sensor OS 500 (right). Pictures are courtesy of OptoSurf GmbH, Ettlingen, Germany

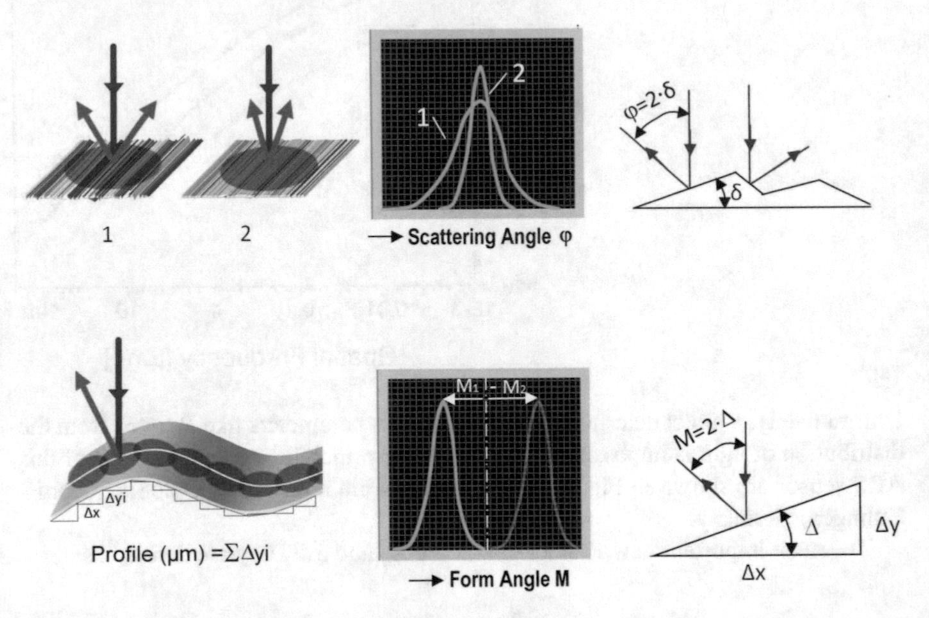

Fig. 5.69 Correlation between surface roughness and A_q (top) and between surface inclination and form angle M (bottom). (Courtesy of Optosurf GmbH, Ettlingen, Germany)

with k being a normalization factor. M is the mean value of the scattered light distribution. A_q corresponds to the variance of the scattered light distribution. M is also a measure for the slope angle Δ of the surface if it is curved within the illumination spot. It is $M = 2 \cdot \Delta$. The correlation of A_q and M with the roughness and form of the workpiece is illustrated in Fig. 5.69. The picture is courtesy of OptoSurf GmbH, Ettlingen, Germany.

In addition to roughness characteristics macroscopic form deviations like roundness PV, waviness, and straightness can be derived. They are derived from the shift of the distribution of scattered light on the detector. For technical surfaces the A_q value correlates with the profile angle (distribution of slopes) Rd_q ($(Rd_q)^2 = A_q$).

The advantages of this ARS sensor are the insensitiveness on distance variations (up to 2 mm for planar surfaces) and vibrations as well as the high measuring rate (up to 2 kHz) and its robustness. The vertical resolution is in the order of 10 nm. The lateral resolution is determined by the measuring spot size which can currently amount to 30 μm, 300 μm, 900 μm, and 7000 μm. The VDA Guideline 2009 from the German automotive association (VDA) refers to this light scattering method using scatterometers similar to the design shown above in Fig. 5.68.

5.9.3 Speckle Based Roughness Determination

When using coherent monochromatic light for illumination of a rough surface an effect additional to specular reflection and scattering can be observed, the formation of speckles and granular speckle patterns. These spatial intensity patterns result from the interferencc of light from object points with random height distribution. Figure 5.70 shows the principle of speckle formation and a typical speckle pattern.

Speckles can be divided into two categories, the *objective speckles* and the *subjective speckles*. Objective speckles are far field speckles that can be directly observed on a screen in the far field. For these speckles many scattered rays contribute. Their mean size depends upon on the ratio of the distance of the detector to the rough surface and the size of the illumination spot. In contrast, for subjective speckles the scattered light is imaged on the screen by a lens. Then, the speckles are formed only by those rays that can pass the lens aperture. Their size is approximately half the Airy disk on the detector.

Speckle based roughness measurements have proven to be a pretty good alternative to TIS and ARS. Since the 1980s speckles are used to determine surface roughness, see e.g. [186–190]. For evaluation of speckle patterns there are two domains with different evaluation methods:

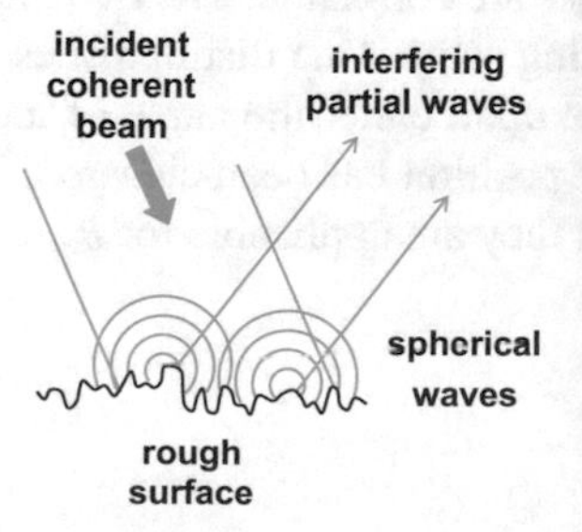

Fig. 5.70 Formation of speckles (left) and typical speckle pattern (right)

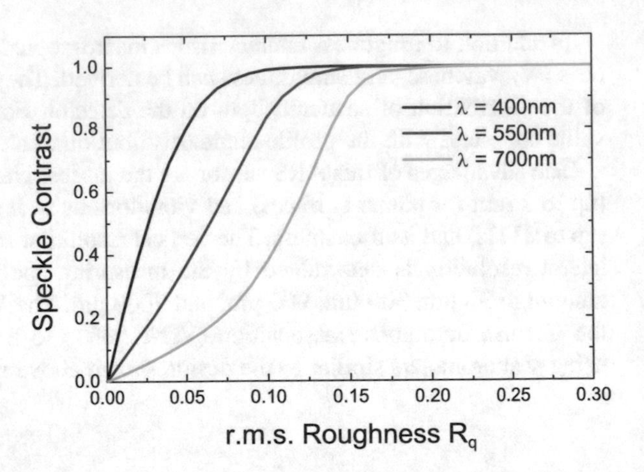

Fig. 5.71 Speckle contrast curves for three different wavelengths

- Speckle contrast evaluation,
- Speckle correlation evaluation.

The *speckle contrast* method uses the intensity contrast C in the speckle image

$$C = \frac{\sigma}{I_{mean}} = \frac{\sqrt{(I_{mean} - I)^2}}{I_{mean}} \tag{5.56}$$

where σ is the standard deviation of the intensity values and I_{mean} is the mean intensity in the image. The contrast C exhibits a characteristic course from $C = 0$ to $C = 1$ versus the roughness parameter R_q. The course is different for different wavelengths. In Fig. 5.71 contrast curves for three different wavelengths are shown. As the speckle contrast runs into saturation this method is only applicable for $R_q < \lambda/4$.

In the *speckle correlation* method the cross correlation coefficient γ_{12} is the relevant quantity.

$$\gamma_{12} = \frac{\left(I_{mean,1} - I_1\right) \cdot \left(I_{mean,2} - I_2\right)}{\sigma_1 \cdot \sigma_2} = \frac{(I_1 \cdot I_2)_{mean} - I_{mean,1} \cdot I_{mean,2}}{\sigma_1 \cdot \sigma_2} \tag{5.57}$$

Here, two speckle images are correlated. The two images differ due to a change in a parameter of the measuring setup. One distinguishes angular, spectral, and spatial correlations in dependence upon either the angle of incidence or the wavelength of incident light or the image position has been changed. The advantage of the speckle correlation methods is that they are applicable for $R_q < 3$–5 µm. They work however only when the speckle contrast is $C = 1$. Figure 5.72 shows the characteristic curves of the cross correlation coefficient γ_{12} versus R_q in spectral correlation method for spectral correlation with four wavelength differences.

If speckles have partially developed another method can be used, the method of *partially developed speckles* [191]. In this method a new quantity *optical roughness* R_{opt} can be retrieved from the autocorrelation function of the intensity distribution of a speckle image. R_{opt} behaves similar to the speckle contrast, i.e. it also runs into saturation with increasing surface roughness parameter and hence is applicable only for smooth surfaces with $R_q < \lambda/4$. Within the funding program Photonics Research Germany by the German Federal Ministry of Education and Research, contract number 13 N13535, a network of the Bremen Institute for Metrology, Automation and Quality Service (BIMAQ), Bremen, and the companies Cosynth GmbH, Oldenburg, and FRT GmbH, Bergisch Gladbach, developed a prototype of a fast speckle-based sensor for the inline roughness determination. The sensor takes and evaluates up to 600 images per second with a field of view of 10 mm in diameter. Figure 5.73 shows a schematic drawing of the roughness sensor and scattered light patterns of surfaces.

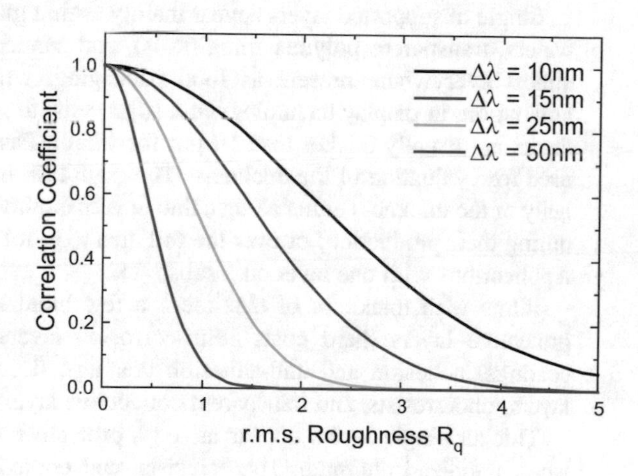

Fig. 5.72 Cross correlation coefficient in spectral correlation method for four different wavelength differences

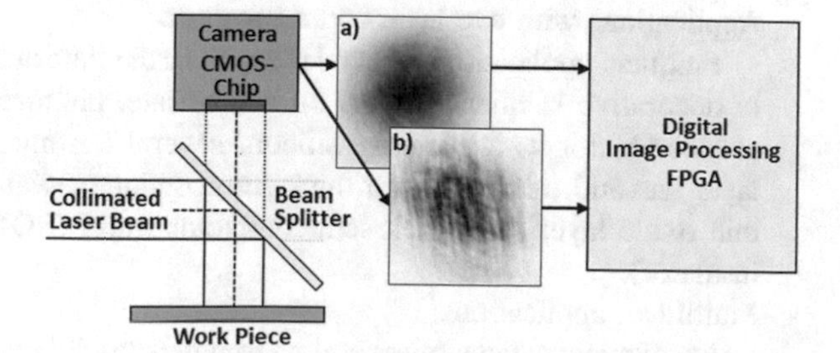

Fig. 5.73 Scheme of the speckle-based optical roughness sensor and scattered light patterns of surfaces with a) $S_a = 3$ nm and b) $S_a = 12.5$ nm. Courtesy of BIMAQ, Bremen, Germany

5.10 Spectral Analysis and Characterization

Surface metrology works with tactile sensors, AFM, optical projection methods, microscopes, and interferometers. These methods are mostly not appropriate if the surface is functionalized by thin coatings or molecular layers as they cannot determine the thickness and other properties of thin films if the film thickness is below a certain limit. The reason is that the modulation of the recorded intensity by a thin film may be less than necessary for the above mentioned methods to recognize the film. Then, other methods must be used which are supplementary to the methods in surface metrology.

In relation to surface metrology mainly reflectometry and spectral ellipsometry using high-valued miniaturized spectrometer modules are the relevant techniques that fulfill the conditions in a production process as well as in laboratories. They will be discussed in this Section.

Thin films of transparent materials play an important role in many fields of technical applications. The applications can be classified into.

- Applications with a single unsupported layer.

  Single unsupported layers appear mainly as thin glass and sapphire sheets and wafers, transparent polymer films (foils), and semiconductor wafers. They are almost everywhere present as food packaging, wrapping, foils, membranes, lamination, in display technology and solar cells, to give some examples. These layers are usually thicker than 10 μm for what a Fast Fourier transform can be used for evaluation of the thickness. The main task is the check of the homogeneity of the thickness either along a line (e.g. in quality control of transparent foils during their production) or over the full area (e.g. for semiconductor wafers).
- Applications with one layer on a substrate.

  Films with thickness of maximum a few hundred nanometers are used as protective layers (hard coats), anti-corrosion layers, broadband antireflection coatings, adhesion and anti-adhesion coatings, decorative coatings, absorbing layers, photoresists, and transparent conductive layers (TCF, TCO).

  Thicker single layers appear as, e.g., protective varnishes (hard coats), finishes, anodized aluminum, photoresists, and epitaxially grown semiconductor layers on semiconductors.
- Applications with two layers on a substrate.

  Frequent applications of two layers on a substrate are: hard coat on a protective or decorative lacquer (primer) on a substrate, photoresist on silica on a wafer, semiconductor wafer (mostly silicon, several ten microns) bonded with a glue layer (several microns) on a thick semiconductor wafer, thin silicon layer on a thin oxide layer on a thick semiconductor wafer (SOI-wafer, SOI = silicon on insulator).
- Multilayer applications.

  Multilayer systems of several nanometers thick layers are often used for high-reflective (HR) and anti-reflective (AR) coatings with 2–8 layers, beam splitter coatings with 4–15 layers, dielectric mirrors with 10 to more than 40 layers, optical filters with 40 to more than 100 layers, and low-E coatings with 2–6 layers including thin Ag layers. Typically, in these stacks layers of low refractive index materials (e.g. MgF_2, SiO_2) are alternating with layers of high refractive index materials (e.g. Al_2O_3, Ta_2O_5, TiO_2, ZrO_2). The thickness of each layer can thoroughly be calculated, respectively the complete layer stack can be optimized.

Other examples for multilayer systems are thin film solar cells and OLEDs (organic light emitting diodes).

- Other applications

  There are several applications where the assumption of thin layers or films is advantagous for the modeling of the optical properties of the systems. One example is the measurement of critical dimensions of vias and trenches in semiconductor wafers.

5.10.1 *Reflectometry*

For a *reflectometric measurement* a light source, a fiber, and optionally a measuring head are commonly used to illuminate the sample with unpolarized light. The reflected light gets collected with the measuring head (optionally) and a second fiber which is connected to a high-valued miniaturized spectrometer. The direction of incidence may include an angle α with respect to the normal on the sample but usually this angle is $\alpha = 0°$ (normal incidence). The principal setup for a reflectometric measurement is sketched in Fig. 5.74a and in Fig. 5.74b the setup of a commercially available spectrometer system MCS 600 with miniaturized diode line spectrometer and light source is shown (picture courtesy of Carl Zeiss Spectrometry GmbH, Jena, Germany). This setup is robust and can easily be adjusted to the ambient conditions at the production site. Moreover, the measured results are stable and reproducible for a long time.

Typically a so-called Y-fiber is used where two separate fibers are assembled so that the branch for the illumination and the branch for the detection of the reflected light are merged in a common branch. Hence, illumination and detection are close together in front of the sample. Without a measuring head the light leaving the fiber gets spread on the sample according to the numerical aperture NA of the fiber (typically $NA = 0.22$). Vice versa the detection fiber only collects the reflected light that enters its aperture. Therefore, the size of the detection spot is determined by the core of the detection fiber as from the widespread illumination spot almost only light can enter the aperture of the detection fiber that is reflected nearly normal to the sample. A measuring head with reproduction scale of 1:1 conserves the light flux and enables larger working distances.

The recorded intensity signal is determined by the reflectivity of the sample, by the spectral distribution of the light source, the spectral sensitivity of the detector,

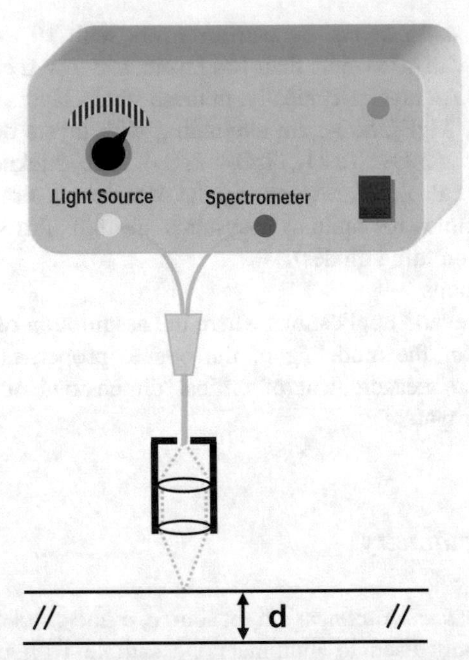

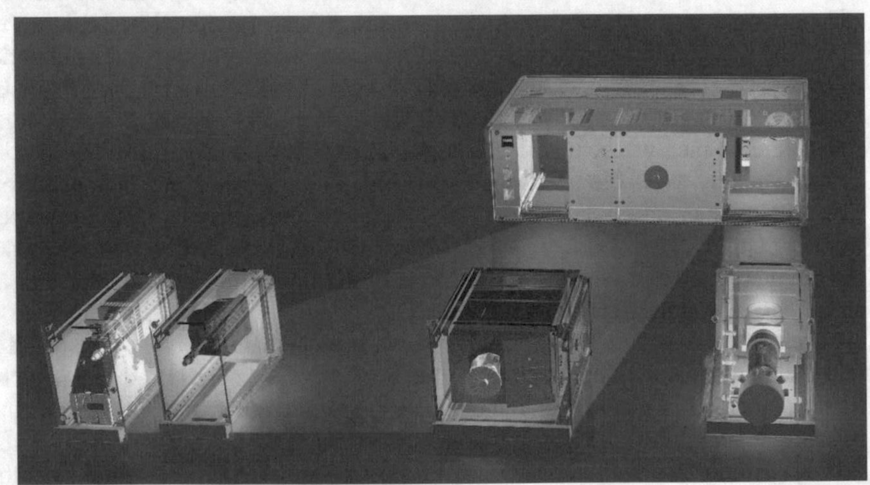

Fig. 5.74 (**a**) Sketch of the principle setup for reflectometric measurement with a miniaturized spectrometer, (**b**) Picture of the setup of a MCS 600 spectrometer system. (Courtesy of Carl Zeiss Spectrometry GmbH, Jena, Germany)

and the transfer characteristics of all optical components. To determine only the sample reflectivity R_{sample} a reference measurement on a reference sample with known reflectivity R_{ref} is necessary. Then, the sample reflectivity R_{sample} is given as

$$R_{sample}(\lambda) = \frac{I(\lambda) - D(\lambda)}{I_{ref}(\lambda) - D(\lambda)} \cdot R_{ref}(\lambda). \quad (5.58)$$

where $I(\lambda)$ and $I_{ref}(\lambda)$ are the measured intensity signals of the sample and the reference sample. They must be corrected for the dark current signal $D(\lambda)$ measured without illumination.

Spectral reflectance measurement can be used to determine the thickness of a single layer or of multiple layers in a stack. As the reflectance of such layer systems depends also on the optical constants of the layer materials, it can also be used to determine optical constants of a single layer. This is an elaborated task that is neither subject of this Section nor of this book but can be read in more detail, e.g. in [192]. Beyond the layer thickness determination also geometric dimensions can be retrieved from reflectance measurements. Both applications are described in the following two sections.

5.10.1.1 Optical Film Thickness Determination

Without loss of generality only the case of a single layer of thickness d and generally complex refractive index $n_1 + i\kappa_1$ on a substrate with generally complex refractive index $n_2 + i\kappa_2$ can be considered for simplicity. A more detailed description for multilayer systems can be read, e.g. in [193]. A parallel beam of incident light with an angle of incidence α hits the top surface S_1 between the layer and the front medium with real refractive index n_0. This is illustrated in Fig. 5.75. The actual magnitude and phase of the incident wave at the boundary between the front medium 0 and the layer medium 1 are not of interest and can be set to the value A_0. A part B_1 of the incident wave gets reflected at the surface S_1 according to the reflection law. The remainder gets refracted into the layer according to Snell's law. It propagates through the layer and hits the surface S_2 between the layer and the substrate under an angle β. Compared to the incident beam at surface S_1 this beam has a phase factor $\exp\left(i\frac{2\pi}{\lambda}\frac{n_1(\lambda)\cdot d}{\cos(\beta)}\right)$. Again, a part of this beam gets reflected and reaches the surface S_1 with the additional phase factor $\exp\left(i\frac{2\pi}{\lambda}\frac{n_1(\lambda)\cdot d}{\cos(\beta)}\right)$, but with modified amplitude.

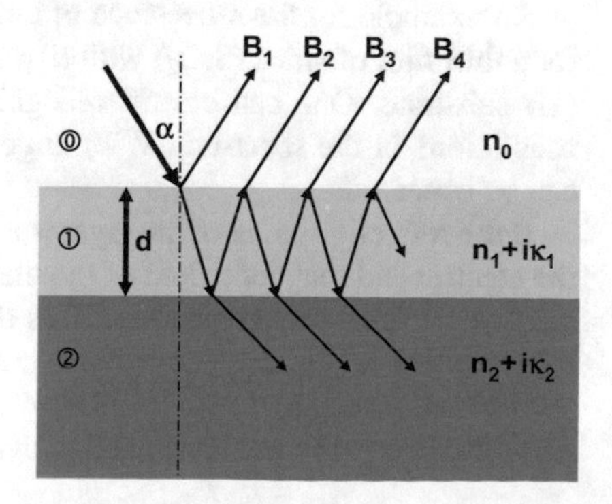

Fig. 5.75 Reflection at a thin layer with refractive index $n_1 + i\kappa_1$ on a substrate with refractive index $n_2 + i\kappa_2$

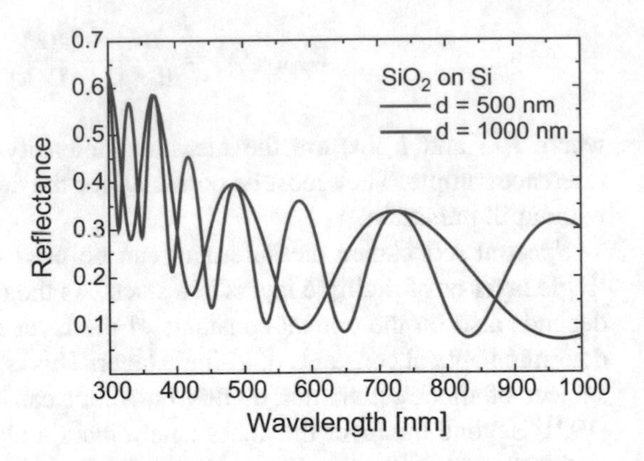

Fig. 5.76 Example for the reflectance of a thin film of silica SiO_2 ($d = 500$ nm and $d = 1000$ nm) on silicon substrate

A part of it gets transmitted and leaves the sample as beam B_2. As on both surfaces reflection and transmission of the corresponding incoming beam occurs, multiple beams B_n occur that all leave the sample in the same direction as the primary reflected beam B_1 but with decreased amplitude and with a phase factor

$$B_{m+1} = t_{10} r_{12} t_{01} (r_{10} r_{12})^{(m-1)} A_0 \exp\left(i \frac{4\pi}{\lambda} \frac{n_1(\lambda) \cdot m \cdot d}{\cos(\beta)} \right) \quad (5.59)$$

with m being a positive integer number $m \geq 1$, and r_{qp} and t_{qp} the Fresnel coefficients for reflection and transmission at the interface (q, p).

For an angle of incidence $\alpha = 0$ the linear superposition of all reflected beams yields for the reflectance $R(\lambda, d)$

$$R(\lambda, d) = \frac{R_{01}(\lambda) + R_{12}(\lambda) + 2\sqrt{R_{01}(\lambda) \cdot R_{12}(\lambda)} \cdot \cos\left(\frac{4\pi}{\lambda} n_1(\lambda) \cdot d\right)}{1 + R_{01}(\lambda) \cdot R_{12}(\lambda) + 2\sqrt{R_{01}(\lambda) \cdot R_{12}(\lambda)} \cdot \cos\left(\frac{4\pi}{\lambda} n_1(\lambda) \cdot d\right)}. \quad (5.60)$$

Obviously, the reflectance is not only determined by the reflectivities at the different interfaces but also by the layer thickness d that is contained only in the oscillating cosine term.

An example for the reflectance of a thin film on a substrate is given in Fig. 5.76 for a thin film of silica (SiO_2) with $d = 500$ nm and with $d = 1000$ nm on a silicon (Si) substrate. One can clearly recognize the layer thickness interference in the oscillations in the spectrum. With larger thickness more periods of the oscillation can be observed.

If the refractive index of the layer is a complex number $n_1 + i\kappa_1$ the magnitudes of the electric and magnetic field of the electromagnetic wave that propagates through the layer decrease by absorption. This decrease also influences the thickness determination as now the magnitude of the oscillating term gets strongly decreased by the absorption. This may cause problems for the detection of the oscillating term if its amplitude becomes smaller than the *signal-to-noise ratio SNR* of the used detector.

Table 5.3 Maximum thickness d_m of an unsupported absorbing layer of different materials for a signal-to-noise ratio $SNR = 1000$. Optical constants of the materials from [194–196]

	Silicon (Si)		Germanium (Ge)		ITO		Aluminum (Al)	
λ[nm]	κ	d_m [μm]	κ	d_m [μm]	κ	d_m [μm]	κ	d_m [μm]
350	2.9911	0.072	2.7040	0.078	0.04496	3.90	4.239	0.054
400	0.3649	0.66	2.2150	0.108	0.03774	5.21	4.861	0.054
500	0.06978	4.19	2.3990	0.126	0.03287	7.26	6.08	0.054
600	0.02586	13.4	1.3667	0.267	0.03479	8.02	7.26	0.054
700	0.009429	42.6	0.4670	0.894	0.04181	7.56	8.31	0.055
800	0.003843	118.3	0.3209	1.47	0.05417	6.43	8.45	0.062
900	0.001847	277.7	0.1851	2.86	0.07309	5.10	8.30	0.070
1100	0.00060759	1028	0.1088	5.93	0.14202	2.64	10.875	0.066
1300	0.00041611	1770	0.07812	9.73	0.30788	1.58	13.147	0.065
1500	0.00035245	2408	0.02568	34.1	0.73963	1.11	15.400	0.064
1700	0.00030487	3153	0.002013	492.3	1.238	0.827	17.567	0.064

The maximum thickness d_m that can then still be determined strongly depends upon the absorption index κ_1 of the film and the *SNR*. It can be estimated to

$$d_m = \frac{\lambda}{4\pi\kappa_1(\lambda)} \cdot \log\left(4 \cdot SNR \cdot \sqrt{R_{01}(\lambda) \cdot R_{12}(\lambda)}\right). \tag{5.61}$$

Here, the cosine term is approximated by the factor 2 (the difference between maximum and minimum value of the cosine). Approximately, d_m is given by

$$d_m \approx (0.4 \quad - \quad 0.75)\frac{\lambda}{\kappa_1(\lambda)} \tag{5.62}$$

for almost all materials and for *SNR* between 1000 and 5000. In the following Table 5.3 the maximum thickness d_m is given for a signal-to-noise ratio of $SNR = 1000$ for different materials at various wavelengths. For calculation of the values in the Table an unsupported layer is assumed for what $R_{12} = R_{01}$. The optical constants of the materials were taken from [194–196].

The example of aluminum demonstrates that for a metal the maximum thickness is less than 100 nm. This is true for all metals. Silicon and germanium become more transparent with increasing wavelength but the onset of the transparency is shifted to longer wavelengths for germanium because of the higher intrinsic absorption. For optical thickness determination of silicon wafers with typical dimensions between 100 and 800 μm it seems appropriate to use the near infrared spectral region above 1100 nm wavelength. The tin doped indium oxide ITO exhibits a window of small absorption in the visible spectral range between 400 and 900 nm but the absorption is still as high as only films of maximum thickness $d_m \approx 5$–8 μm can be measured optically.

The evaluation of the thickness of the layer can be made by two methods: a *Fast Fourier Transformation* (FFT) analysis or a *nonlinear regression analysis* of the measured reflectance (or transmittance) spectrum.

For applying a FFT analysis it is necessary that at least one full oscillation is in the measured spectrum. If the spectral range of the spectrometer is given by λ_{min} and λ_{max}, the first complete oscillation in the spectrum of a film with refractive index $n(\lambda)$ can be recognized if one maximum appears at λ_{min} and the second maximum appears at λ_{max}. Then, the minimum thickness available from FFT is

$$\mathrm{d_{min}} = \frac{1}{2 \cdot \left(\frac{\mathrm{n}(\lambda_{\mathrm{min}})}{\lambda_{\mathrm{min}}} - \frac{\mathrm{n}(\lambda_{\mathrm{max}})}{\lambda_{\mathrm{max}}} \right)}. \tag{5.63}$$

It is in the order of 200 nm for a spectral range from 360 to 1000 nm and a typical refractive index of $n = 1.5$. It increases with decreasing wavelength range.

All evaluable thickness values are integer multiples of the smallest thickness. That means, if the actual thickness of the film does not match one of these discrete results, the resulting power spectrum from FFT exhibits leakage in addition to the leakage caused by the finite wavelength interval and the obtained thickness is only close to the actual thickness. Due to discreteness of the obtainable thickness results the uncertainty in FFT analysis amounts at best to the half of the minimal evaluable thickness. Within these uncertainty bounds the obtained thickness is reliable. A thickness larger than the maximum evaluable thickness $d_{max} = (M - 1) \cdot d_{min}$ with M = number of sampling points (wavelengths), will lead to oscillations that cannot be resolved in the reflectance spectrum.

For the reflectance spectra shown in Fig. 5.76 the minimal evaluable thickness amounts to 133 nm. Hence, thickness values can be expected from FFT analysis that are multiples of 133 nm with an uncertainty of 67 nm. In Fig. 5.77 the power spectral distribution of the two reflectance spectra in Fig. 5.76 are shown. The results for the thickness from FFT analysis are d = 532 nm and d = 1065 nm corresponding to four times and eight times the minimal evaluable thickness. The small peak at pixel no. 15 corresponds to the second harmonic of the peak at pixel no. 8.

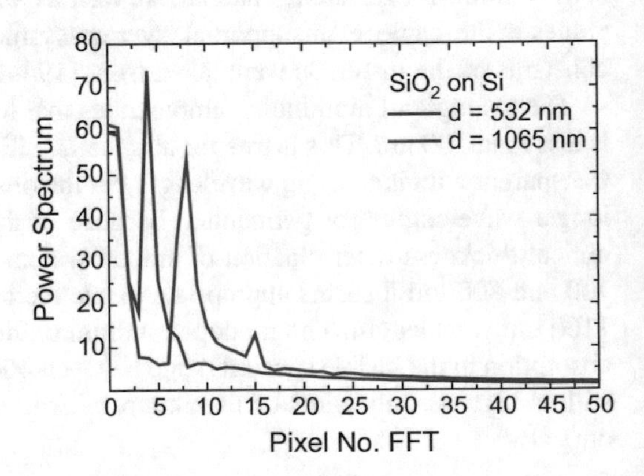

Fig. 5.77 Power spectral distribution of the reflectance spectra in Fig. 5.76. The peak at pixel no. 0 is artificially decreased

In practice, often the refractive indices in Eq. (5.63) are replaced by a single mean refractive index. Then, errors of up to 5% are possible for materials with high refractive index like semiconductors. A better approach is to use an *effective refractive index* n_{eff} which is always higher than the refractive indices of the material in the considered wavelength range. The problem with n_{eff} is that it is usually unknown. If the refractive indices $n(\lambda_{min})$ and $n(\lambda_{max})$ at the borders of the wavelength range are known, it can be calculated from

$$n_{eff} = \frac{\left(\frac{n(\lambda_{min})}{\lambda_{min}} - \frac{n(\lambda_{max})}{\lambda_{max}}\right)}{\left(\frac{1}{\lambda_{min}} - \frac{1}{\lambda_{max}}\right)}. \tag{5.64}$$

In summary, if applicable, FFT is a very fast method to determine film thickness but its resolution is limited due to the discreteness of the sampling points and the finite interval of sampling points. For more details on FFT applied on reflectance spectra of thin films the reader is referred to [193].

The *regression analysis* is more complicated than the FFT as it uses the calculation of the reflectance spectrum of a layer stack in the whole spectral range. On the other hand, it is well suited for small layer thickness down to a few 10 nm depending upon the layer material and yields more precise results. In principle, it can be applied also on films with larger thickness but the computational effort drastically increases with increasing thickness. A reasonable (but not fixed) upper application limit is $d \approx 10$ μm.

The regression analysis used for layer thickness determination fits a model that depends upon the thickness and the optical constants of the layer (e.g. Eq. (5.60)) to a given set of measured data (spectral reflectance). The best figure of merit for the goodness of a fit algorithm is the quadratic deviation χ^2

$$\chi^2 = \sum_{m=1}^{M} \left(\frac{R(\lambda_m) - R_{calc}(\lambda_m, d_1, \ldots, d_N)}{\sigma_m} \right)^2 \tag{5.65}$$

The value σ_m is the standard deviation of the *m*-th measurement value. Often, these standard deviations are set to a constant number. The used fit algorithm must now provide a technique to minimize χ^2 iteratively to find the next set of thickness parameters. If all parameters from the $(k + 1)$-th iteration step deviates relatively from that from the *k*-th iteration step less than a given accuracy the iteration stops. Then, a set of thickness parameters of the *N* layers in a layerstack is found for which the calculated spectrum deviates minimal from the measurement. Two algorithms are well-known that achieve this: the *Levenberg-Marquardt algorithm* [197, 198] that uses first order derivatives of χ^2 versus the parameters for improvement of the result in the next iteration step, and the *downhill simplex algorithm* according to Nelder and Mead [199] that tries to minimize the *N*-dimensional volume spanned by $N + 1$ χ^2-values. Both methods can be read more detailed in the book *Numerical*

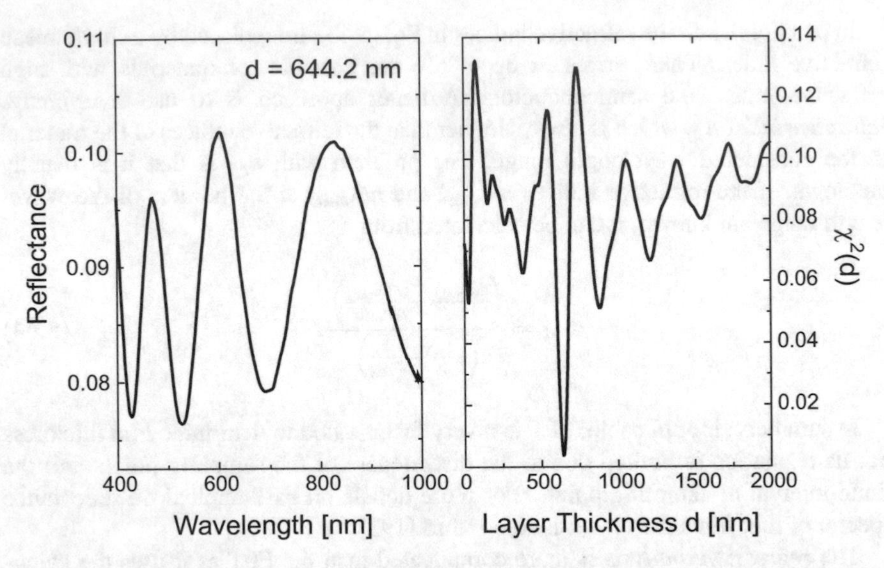

Fig. 5.78 Measured reflectance of a photoresist film with $d = 644.2$ nm on glass substrate (left) and the calculated size dependence of the quadratic deviation for this sample (right)

Recipes – The Art of Scientific Computing [200]. Note that the actual value of χ^2 is meaningless. It is of major importance that it becomes minimal.

Once found the optimal set of thickness parameters their uncertainty can be calculated from the normal equations of the least-squares problem (see again [200]). The covariance matrix is closely related to the standard uncertainties of the obtained thickness parameters. The diagonal elements correspond to variances (squared uncertainties) of the parameters.

For the regression analysis the resolution can be estimated from Eq. (5.60) when resolving first for the thickness d

$$d = \frac{\lambda}{4\pi \cdot n(\lambda)} \cos^{-1}\left(\frac{R(\lambda) \cdot (1 + R_{01}(\lambda) \cdot R_{12}(\lambda)) - R_{01}(\lambda) - R_{12}(\lambda)}{2 \cdot \sqrt{R_{01}(\lambda) \cdot R_{12}(\lambda)} \cdot (1 - R(\lambda))}\right) \quad (5.66)$$

As the arcus cosine function cannot exceed 2π it can be replaced by 2π for estimation of the upper resolution limit. Then, the resolution becomes

$$\Delta d \leq \frac{\Delta\lambda}{2 \cdot n(\lambda)} \approx \frac{3 \cdot (\lambda_{max} - \lambda_{min})}{2 \cdot n(\lambda_{max}) \cdot N} \quad (5.67)$$

The factor 3 takes into account that the resolution according to the Rayleigh criterion is approximately three times the pixel resolution.

There are two main problems when using regression analysis that must be considered in more detail. The first problem is with the initial values. It is demonstrated for a photoresist layer on a glass substrate. In Fig. 5.78 the measured

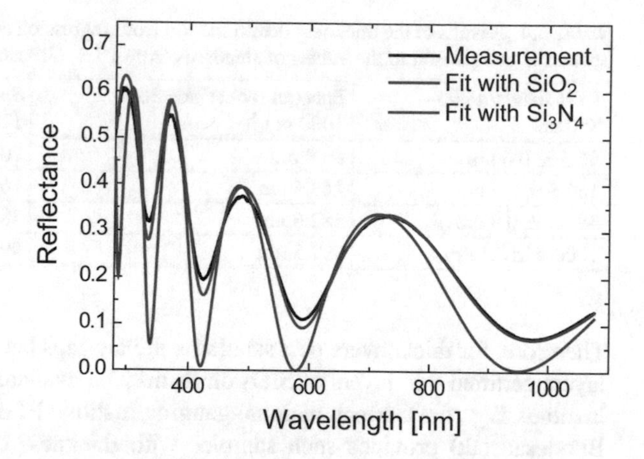

Fig. 5.79 Measured reflectance spectrum of a 500 nm thick film of SiO_2 on Si (black) and the results of the regression analysis using refractive index data of SiO_2 (blue) and of Si_3N_4 (red) for the film

reflectance spectrum of a photoresist film with a determined thickness of $d = 644.2$ nm is shown at left and at right the dependence of χ^2 on the thickness d for this sample.

It becomes obvious that χ^2 has several local minima. One of them (the deepest) yields the film thickness of $d = 644.2$ nm. However, it is only achieved if the initial value of d is chosen in the interval [520 nm, 715 nm]. Starting with a value in the interval [400 nm, 520 nm] leads to $d \approx 384$ nm, and starting with a value in the interval [715 nm, 850 nm] leads to $d \approx 884$ nm. These values belong to the next neigboring local minima in χ^2. For more layers in the stack the intervals for the initial values become the smaller the more layers with unknown thickness are in the stack. Hence, a logical consequence is that the nonlinear regression yields unambiguous results only for a small number of layers. In fact, for more than three layers in the stack with unknown thickness there always exist more than one solution with excellent matching of fit and measurement.

The next problem arises with the refractive index data of the film material. Using inappropriate data will lead to a larger quadratic deviation on the one hand and to a wrong thickness on the other hand. The thickness becomes lower if the refractive index is higher than the appropriate value and vice versa. This is illustrated exemplarily in Fig. 5.79 for a thin film of SiO_2 on Si substrate using optical constants of silica SiO_2 (blue line) and using optical constants of silicon nitride Si_3N_4 (red line) in the fit.

As expected the fit with optical constants of SiO_2 yield the best fit with $d = 502.8$ nm and $\chi^2 = 0.386$. Optical constants of Si_3N_4 yield a best fit with $d = 355.5$ nm and $\chi^2 = 13.255$. The use of the high refractive Si_3N_4 yields a smaller size by a factor that approximately corresponds to the ratio of the refractive indices of Si_3N_4 and SiO_2. Beyond this, clearer deviations from the measurement are obvious which is also reflected in the high χ^2.

In practice, it is difficult to find certified standards for thickness determination for thick layers where FFT can be applied as well as for thin layers. The reason is that besides the thickness also the refractive index must be well-known and certified.

Table 5.4 Results of the thickness determination from regression analysis of measured reflectance spectra in comparison to the values of standards from PTB, Germany

PTB standard 300–850 nm	Spectral reflectance 360–1000 nm	Spectral reflectance 900–1700 nm
66.5 ± 0.6 nm	66.8 nm	70.3 nm
160.8 ± 0.7 nm	163.6 nm	160.9 nm
381.5 ± 0.9 nm	382.6 nm	383.3 nm
1000.4 ± 2.1 nm	997.5 nm	998.3 nm

Therefore, for thick layers best standards are air gaps between optical flats. For thin layers certified thin layers of SiO_2 on Si may be available from a national gauging institute. E.g., the German national gauging institute PTB (Physikalisch Technische Bundesanstalt) provides such samples with thickness of the SiO_2-layer between approximately 50 nm and 1000 nm. Table 5.4 comprises results of reflectance measurements with two spectrometers with different wavelength range on such certified samples. The values from PTB include the information on the expanded uncertainty with coverage factor $k = 2$.The differences between the obtained thickness values and the certified values may be caused by different reasons:

- The best fit to the measured spectrum must fit simultaneously at many sampling points. If the number of sampling points differs from that in the certification process this can result in another thickness than the certified one.
- The errors in the used optical constants will affect the thickness determination as well other sets of optical constants than that used for specification can lead to different results.
- When using a spectral range that not includes the wavelengths used for the certification of the standard, problems may arise because the (n, κ)-data are perhaps only partly available. Moreover, for longer wavelengths the minimal detectable thickness will shift to larger sizes.

Anyhow, the mean value of a series of repeated measurements can have an excellent standard deviation for repeatability and reproducibility with values less than 1 nm.

5.10.1.2 Critical Dimensions Determination

Spectral reflectometry can also be used for the determination of topographic values. In particular critical dimensions of through silicon vias (TSV) and trenches can be determined from the reflectance spectrum. The critical dimensions of a TSV or a trench are the depth, the top diameter, the bottom diameter, and, if top and bottom diameter differ, the taper angle. Usually, the silicon wafer has a coating of SiO_2 in the order of 1–2 μm on top surface. For process control TSV arrays with fixed pitch are used.

The established method uses infrared reflectometry or ellipsometry for optical measurement and modeling the measured reflectance using EMA (Effective Medium Approximation) models. This method is called *Model Based Infrared Reflectometry* (MBIR). EMA models replace the inhomogeneous matter of two (or more) nonmixable components by a homogeneous matter with effective optical constants. For calculation of the effective optical constants the optical constants of the components (voids, silicon) and the volume fraction or filling factor ϕ of the voids must be known. The latter can easily be determined in a regular array with known pitches in x- and y- direction and contains the volume of the single void or trench. Then, one of the most popular EMA models that consider explicitely the shape of the inclusions is used to calculate the dielectric function of the effective medium. The most popular EMA models stem from J. C. M. Garnett [201].

$$\frac{\varepsilon_{\mathrm{eff}}-\varepsilon_{\mathrm{M}}}{\varepsilon_{\mathrm{eff}}+(\mathrm{D}-1)\cdot\varepsilon_{\mathrm{M}}}=\phi\frac{\varepsilon-\varepsilon_{\mathrm{M}}}{\varepsilon+(\mathrm{D}-1)\cdot\varepsilon_{\mathrm{M}}},\tag{5.68}$$

well known as Maxwell-Garnett theory, and from D. A. G. Bruggeman [202].

$$\phi\cdot\frac{\varepsilon-\varepsilon_{\mathrm{eff}}}{\varepsilon+(\mathrm{D}-1)\cdot\varepsilon_{\mathrm{eff}}}+(1-\phi)\cdot\frac{\varepsilon_{\mathrm{M}}-\varepsilon_{\mathrm{eff}}}{\varepsilon_{\mathrm{M}}+(\mathrm{D}-1)\cdot\varepsilon_{\mathrm{eff}}}=0.\tag{5.69}$$

The original 3D formulas are obtained by inserting $D = 3$ for the dimension parameter D. For $D = 2$ one obtains the 2D solutions.

The use of radiation in the near infrared to probe the structures is induced by the fact that at wavelengths $\lambda > 1100$ nm silicon microstructures become transparent. Then, one obtains an interference pattern in the reflectance spectrum that encodes details of the vias or trenches shape and depth. The general procedure in MBIR is

- Replace the hardcoat layer (Si_3N_4 or SiO_2) with TSVs or trenches by a layer of Si_3N_4 or SiO_2 with a certain amount of voids. The dielectric function of this layer gets calculated with a two-dimensional EMA model.
- Replace the silicon region with vias or trenches by a layer of Si with voids. The effective dielectric function is again calculated with 2D-EMA models. If necessary, this region can be divided into multiple layers to consider the shape of the via or trench more detailed.
- Replace the bottom silicon region with vias or trenches by a multilayer stack of thin layers (graded layer) to consider different depths of the trenches. The optical constants of each layer are calculated from an EMA model.
- Calculate the reflectance of this multilayer stack with a silicon substrate.
- Fit with Levenberg-Marquardt algorithm to determine the critical dimensions and the depth.

5.10.2 Spectroscopic Ellipsometry

Ellipsometry is a technique originally developed by Paul Drude to study polarized light reflected from liquid surfaces and from solids coated with thin films [203]. However, just with the development of electronics and computers this over hundred years old technique became relevant because then it became possible to fit the experimental data to the physics-based first principles equations in a short time. The name *ellipsometry* originated in 1945 in a paper of Alexandre Rothen [204]. Spectral ellipsometry (SE) can analyze complex structures such as multilayers, interface roughness, inhomogeneous layers, anisotropic layers, and much more.

Azzam and Bashara [205] published in 1977 the book *Ellipsometry and Polarized Light* which has become the key source in ellipsometry. Further books that cover the theory of ellipsometry, fundamental principles, the instrumentation, and applications have been published later [206–211].

For an ellipsometric measurement a light source that provides unpolarized light and a polarizer are used to illuminate the sample with a light beam in an accurately known polarization state. Optionally, an optical retarder is placed between the polarizer and the sample. The direction of incidence includes an angle α with respect to the normal on the sample. Specular reflection of the beam from the sample surface leads to an emergent beam in an elliptical polarization state. It trespasses an analyzer and gets detected by an optical detector. Optionally, an optical retarder is placed between the sample and the analyzer. Usually, the analyzer is rotated (Rotating Analyzer Ellipsometry = RAE) to enable at least eight different measurements but it may also be that the polarizer is rotated (Rotating Polarizer Ellipsometry = RPE). The principal setup for an ellipsometric measurement is sketched in Fig. 5.80.

Ellipsometry measures the complex ratio ρ of the Fresnel reflection coefficient of the p-polarized and s-polarized component of the reflected light:

$$\rho = \frac{r_p(\alpha)}{r_s(\alpha)} = \frac{|r_p(\alpha)| \cdot \exp(i \cdot \Delta_p)}{|r_s(\alpha)| \cdot \exp(i \cdot \Delta_s)} = \sqrt{\frac{R_p}{R_s}} \cdot \exp(i\Delta) = \tan(\psi) \cdot \exp(i\Delta) \quad (5.70)$$

where $\tan(\psi)$ is the amplitude ratio and Δ is the phase shift of the p- and s- reflection coefficients. They are the *ellipsometric parameters* often also given as $\tan(\psi)$ and $\cos(\Delta)$ or only as ψ and Δ. Although the spectral ellipsometric measurement delivers two observables, ψ and Δ, bear in mind that the two observables are not independent of each other. They belong to the same complex-valued quantity ρ.

Different measurement techniques of the polarization after reflection exist. For them other components like modulators or compensators can be added. Modern ellipsometer adjust all components automatically and calculate the ellipsometry parameters very fast.

Spectroscopic ellipsometry (SE) measures the change in polarization of light simultaneously at different wavelengths. The commercially available spectral range covers 150 nm to 33 μm but not in one single spectral range. It also allows the determination of the properties (thickness, complex refractive index) of a layer

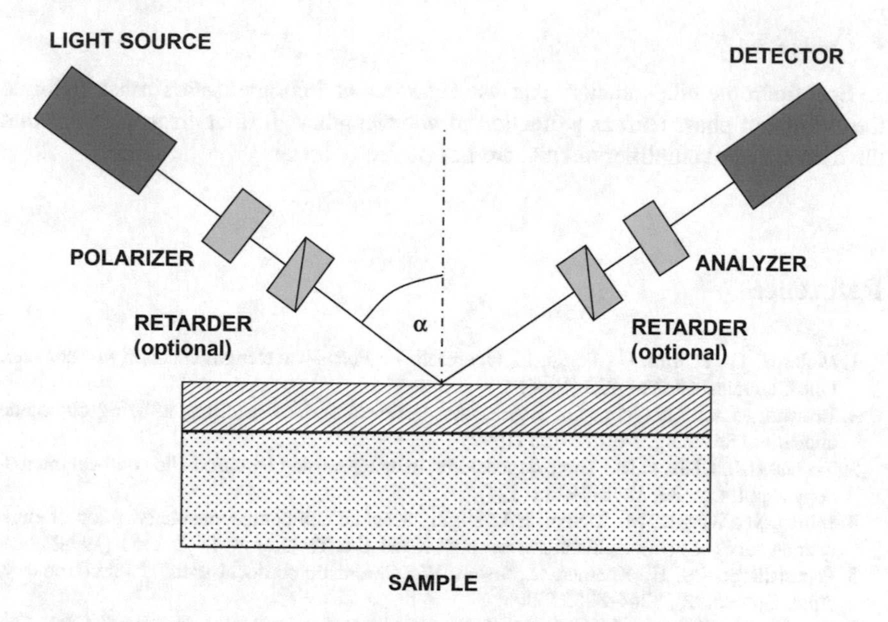

Fig. 5.80 Sketch of an ellipsometric thin film measurement

stack. For that purpose, physical models for the layer stack and the optical constants are necessary similar to reflectometry. A nonlinear regression analysis is used to get the best fit of calculated data to measured data.

If the sample is an ideal bulk the real and imaginary parts of the pseudo complex dielectric function may be calculated from

$$\varepsilon_1 + i\varepsilon_2 = \sin^2(\alpha) \cdot \left[1 + \tan^2(\alpha) \cdot \left(\frac{1-\rho}{1+\rho}\right)\right]. \tag{5.71}$$

with the knowledge of the angle of incidence α and presuming the ambient to be air.

Ellipsometry is not free of errors. Sources of systematic errors are:

- Azimuthal alignment of optical elements

Each optical element must be aligned to ensure a high quality ellipsometric measurement. The azimuthal alignment errors are typically in the order of 0.02° for each element. All alignment errors of the several optical elements add up and may result in a significant systematic error in the experiment.

- The angle of incidence

The angle of incidence is a particularly difficult parameter to measure and its error is hard to quantify. Generally, it is very hard to measure this quantity to better than ≈ 0.02°. Additionally, the used light source is not perfectly collimated so the sample is actually illuminated with a light beam with a distribution of angles of incidence.

- Calibration

Spectroscopic ellipsometers that use retarders and compensators must calibrate the amount of phase shift as a function of wavelength. A further error source is that the detectors and amplifier circuits are not perfectly linear.

References

1. Molesini, G., Pedrini, G., Poggi, P., Quercioli, F.: Focus-wavelength encoded profilometer. Opt. Commun. **49**, 229–233 (1984)
2. Browne, M.A., Akinyemi, O., Boyde, A.: Confocal surface profiling utilizing chromatic aberration. Scanning. **14**, 145–153 (1992)
3. Tiziani, H.J., Uhde, H.M.: Three dimensional imaging sensing by chromatic confocal microscopy. Appl. Opt. **33**, 1838–1843 (1994)
4. Jordan, M., Wegner, M., Tiziani, H.J.: Highly accurate non-contact characterization of engineering surfaces using confocal microscopy. Meas. Sci. Technol. **9**, 1142–1151 (1998)
5. Papastathopoulos, E., Koerner, K., Osten, W.: Chromatic confocal spectral interferometry. Appl. Opt. **45**(32), 8244–8252 (2006)
6. Kunkel, M., Schulze, J.: Noncontact measurement of central lens thickness. Glas. Sci. Technol. **78**(5), 245–247 (2005)
7. Compaan, K., Kramer, P.: The Philips VLP system. Philips Tech. Rev. **33**(7), 178–180 (1973)
8. Bouwhuis, G., Braat, J.: Principles of Optical Disc Systems. Hilger, Bristol (1985)
9. Bricot, C., Lehureau, J.C., Fokussiereinrichtung, German patent DE 2501124 A1 (1974)
10. Bricot, C., Lehureau, F., Puech, C., le Carvennec, F.: Optical readout of videodisc. IEEE Trans. Consum. Electron. **CE-22**(4), 304–308 (1976)
11. Scheimpflug, T.: Improved method and apparatus for the systematic alteration or distortion of plane pictures and images by means of lenses and mirrors for photography and for other purposes, British Patent No. 1196 (1904)
12. Kühmstedt, P., Munkelt, C., Heinze, M., Bräuer-Burchardt, C., Notni, G.: 3D shape measurement with phase correlation based fringe projection. Proc. SPIE. **6616**, 66160B (2007)
13. Wiegmann, A., Wagner, H., Kowarschik, R.: Human face measurements by projection bandlimited random patterns. Opt. Express. **14**, 7692–7968 (2006)
14. Grosse, M., Schaffer, M., Harendt, B., Kowarschik, R.: Fast data acquisition from three-dimensional shape measurements using fixed-pattern projection and temporal coding. Opt. Eng. **50**, 100503 (2011)
15. Schaffer, M., Grosse, M., Kowarschik, R.: High-speed pattern projection from three-dimensional shape measurement using laser speckles. Appl. Opt. **49**, 3622–3629 (2010)
16. Schaffer, M., Grosse, M., Harendt, B., Kowarschik, R.: Coherent two-beam interference fringe projection for high speed three-dimensional shape measurements. Appl. Opt. **52**, 2306–2311 (2013)
17. Abbe, E.: Theorie des Mikroskops und der mikroskopischen Wahrnehmung. Arkiv. Mikroskop. Anat. **9**(1), 413–418 (1873)
18. Berek, M.: Grundlagen der Tiefenwahrnehmung im Mikroskop mit einem Anhang über die Bestimmung der obersten Grenze des unvermeidlichen Fehlers einer Messung aus der Häufigkeitsverteilung des zufälligen Maximalfehler. Sitzungsberichte der Gesellschaft zur Beförderung der gesamten Naturwissenschaften zu Marburg. **62**(6), 189–223 (1927)
19. Zernike, F.: Das Phasenkontrastverfahren bei der mikroskopischen Beobachtung. Z. techn. Physik. **16**, 454–457 (1935)
20. Zernike, F.: How I discovered phase contrast. Science. **121**(3141), 345–349 (1955)
21. Normarski, G.: Interféromètre à polarisation, French patent No. 1.059.123 (1952)

22. Nomarski, G.: Microinterféromètre différentiel à ondes polarisées. J. Phys. Radium. **16**, S9–S13 (1955)
23. Minsky, M.: Microscopy Apparatus, U.S. Patent 3, 013, 467 (1961)
24. Minsky, M.: Memoir on inventing the confocal scanning microscopy. Scanning. **10**, 128–138 (1988)
25. Ruprecht, A.K., Wiesendanger, T.F., Tiziani, H.J.: Signal evaluation for high-speed confocal measurements. Appl. Opt. **41**(35), 7410–7415 (2002)
26. Wilson, T. (ed.): Confocal Microscopy. Academic, New York (1990)
27. Pawley, J.B. (ed.): Handbook of Biological Confocal Microscopy. Plenum Press, New York (1995)
28. Corle, T.R., Kino, G.S.: Confocal Scanning Optical Microscopy and Related Imaging Systems. Academic, New York (1996)
29. Sheppard, C.J.R., Shotton, D.M.: Confocal Laser Scanning Microscopy. BIOS Scientific Publishers, Oxford (1997)
30. Paddock, S.W. (ed.): Confocal Microscopy: Methods and Protocols. Humana Press, Totowa (1999)
31. Diaspro, A. (ed.): Confocal and Two-Photon Microscopy: Foundations, Applications, and Advances. Wiley-Liss, New York (2002)
32. Egger, M.D., Petran, M.: New reflected-light microscope for viewing unstained brain and ganglion cells. Science. **157**, 305–307 (1967)
33. Petráň, M., Hadravský, M., Egger, M.D., Galambos, R.: Tandem-scanning reflected-light microscope. J. Opt. Soc. Am. **58**(5), 661–664 (1968)
34. Nipkow, P.: Elektrisches Teleskop, German Patent DE30105 (C), 15 January 1885
35. Tiziani, H.J., Achi, R., Krämer, R.N., Wiegers, L.: Theoretical analysis of confocal microscopy with microlenses. Appl. Opt. **35**(1), 120–125 (1996)
36. Ichihara, A., Tanaami, T., Isozaki, K.: High-speed confocal fluorescent microscopy using a Nipkow scanner with microlenses for 3-D imaging of single fluorescent molecule in real time. Bioimages. **42**, 57–62 (1996)
37. Hell, S.W.: Double-confocal scanning microscope, European Patent EP 0491289, 24 June 1992
38. Hell, S.W., Stelzer, E.H.K.: Properties of a 4π confocal fluorescence microscope. J. Opt. Soc. Am. A. **9**(12), 2159–2166 (1992)
39. Hell, S.W., Stelzer, E.H.K., Lindek, S., Cremer, C.: Confocal microscopy with an increased detection aperture: type-B 4π confocal microscopy. Opt. Lett. **19**(3), 222–224 (1994)
40. Cremer C., Cremer T.: 4Π Punkthologramme: Physikalische Grundlagen und mögliche Anwendungen. Enclosure to Patent application DE 2116521, 12 October 1972
41. Cremer, C., Cremer, T.: Considerations on a laser-scanning-microscope with high resolution and depth of field. Microsc. Acta. **81**(1), 31–44 (1978)
42. Hell, S.W., Wichmann, J.: Breaking the diffraction resolution limit by stimulated emission: stimulated-emission-depletion fluorescence microscopy. Opt. Lett. **19**, 780–782 (1994)
43. Hell, S.W.: Far-field optical nanoscopy. Science. **316**, 1153–1158 (2007)
44. Hell, S.W., Kroug, M.: Ground-state-depletion fluorescence microscopy: a concept for breaking the diffraction resolution limit. Appl. Phys. B Lasers Opt. **60**, 495–497 (1995)
45. Hell, S.W.: Toward fluorescence nanoscopy. Nat. Biotechnol. **21**(11), 1347–1355 (2003)
46. Betzig, E., Patterson, G.H., Sougrat, R., Lindwasser, O.W., Olenych, S., Bonifacino, J.S., Davidson, M.W., Lippincott-Schwartz, J., Hess, H.F.: Imaging intracellular fluorescent proteins at nanometer resolution. Science. **313**, 1642–1645 (2006)
47. Rust, M.J., Bates, M., Zhuang, X.: Sub-diffraction-limit imaging by stochastic optical reconstruction microscopy (STORM). Nat. Methods. **3**, 793–796 (2006)
48. Dertinger, T., Colyer, R., Iyer, G., Weiss, S., Enderlein, J.: Fast, background-free, 3D super-resolution optical fluctuation imaging (SOFI). Proc. Natl. Acad. Sci. U. S. A. **106**, 22287–22292 (2009)

49. Pohl, D.W., Denk, W., Lanz, M.: Optical stethoscopy: image recording with resolution λ/20. Appl. Phys. Lett. **44**, 651–653 (1984)
50. Pohl, D.W., Denk, W., Dürig, U.: Optical stethoscopy: imaging with λ/20. Proc. Soc. Photo-Opt. Instrum. Eng. **565**, 56–61 (1986)
51. Dürig, U., Pohl, D.W., Rohner, F.: Near-field optical scanning microscopy. J. Appl. Phys. **51**, 3318–3327 (1986)
52. Pohl, D.W.: Scanning near-field optical microscopy (SNOM). Adv. Opt. Electron Microsc. **12**, 243–312 (1991)
53. Betzig, E., Trautman, J.K.: Near-field optics: microscopy, spectroscopy, and surface modification beyond the diffraction limit. Science. **257**, 189–195 (1992)
54. Heinzelmann, H., Pohl, D.W.: Scanning near-field optical microscopy. Appl. Phys. A Mater. Sci. Process. **59**, 89–101 (1994)
55. Synge, E.H.: A suggested method for extending the microscopic resolution into the ultramicroscopic region. Phil. Mag. **6**, 356–362 (1928)
56. Ash, E.A., Nicholls, G.: Super-resolution aperture scanning microscope. Nature. **237**(5357), 510–512 (1972)
57. Girard, C., Dereux, A.: Near-field optics theories. Rep. Prog. Phys. **59**, 657–699 (1996)
58. Carré, P.: Installation et utilisation du comparateur photoelectrique et Interferentiel du Bureau International de Poids et Measures. Metrologia. **1**, 13–23 (1966)
59. Crane, R.: Interference phase measurement. Appl. Opt. **8**, 538–542 (1969)
60. Wyant, J.C.: Double frequency grating lateral shear interferometer. Appl. Opt. **12**, 2057–2060 (1973)
61. Bruning, J.H., Herriott, D.R., Gallagher, J.E., Rosenfeld, D.P., White, A.D., Brangaccio, D.J.: Digital wavefront measuring interferometer for testing optical surfaces and lenses. Appl. Opt. **13**, 2693–2703 (1974)
62. Wyant, J.C.: Use of an ac heterodyne lateral shear interferometer with real-time wavefront correction systems. Appl. Opt. **14**, 2622–2626 (1975)
63. Schwider, J., Burow, R., Elssner, K.-E., Grzanna, J., Spolaczyk, R., Merkel, K.: Digital wavefront measuring interferometry: some systematic error sources. Appl. Opt. **22**, 3421–3432 (1983)
64. Schwider, J.: Advanced evaluation techniques. In: Wolf, E. (ed.) Interferometry, in Progress in Optics, vol. XXVIII, pp. 271–359. Elsevier, Amsterdam (1990)
65. Hariharan, P., Oreb, B.F., Eiju, T.: Digitial phase-shifting interferometry: a simple error-compensating phase calculation algorithm. Appl. Opt. **26**, 2504–2506 (1987)
66. Polster, H.D., Pastor, J., Scott, R.M., Crane, R., Langenbeck, P.H., Pilston, R., Steinberg, G.: New developments in interferometry. Appl. Opt. **8**, 521–556 (1969)
67. Wyant, J.C., MacGovern, A.J.: Computer generated holograms for testing aspheric optical elements, Applications de L'Holographie, Laboratoire de Physique Generale et Optique, pp. 13–18. Universite de Besancon, Besancon (1970)
68. MacGovern, A.J., Wyant, J.C.: Computer generated holograms for testing optical elements. Appl. Opt. **10**, 619–624 (1971)
69. Wyant, J.C., Bennett, V.P.: Using computer generated holograms to test aspheric wavefronts. Appl. Opt. **11**, 2833–2839 (1972)
70. Malacara, D., Servin, M., Malacara, Z.: Interferogram analysis for optical testing. Marcel Dekker Inc., New York (1998)
71. Malacara, D., Creath, K., Schmit, J., Wyant, J.C.: Testing of aspheric wavefronts and surfaces. In: Malacara, D. (ed.) Optical Shop Testing, 3rd edn, pp. 477–488. Wiley, New York (2007)
72. Dörband, B., Tiziani, H.J.: Testing aspheric surfaces with computer generated holograms: analysis of adjustment and shape errors. Appl. Opt. **24**, 2604–2611 (1985)
73. J. Schwider, Interferometric tests for aspherics, in fabrication and testing of Aspheres, 24 of OSA Trends in Optics and Photonics, M. Taylor, M Piscotty, A. Lindquist (Optical Society of America, Washington, DC, 1999)., paper T3

74. Pruss, C., Reichelt, S., Tiziani, H.J., Olsen, W.: Computer generated holograms in interferometric testing. Opt. Eng. **43**, 2534–2540 (2004)
75. Zernike, F.: Beugungstheorie des Schneidenverfahrens und Seiner Verbesserten Form, der Phasenkontrastmethode. Physica. **1**(8), 689–704 (1934)
76. Zernike, F.: Diffraction theory of knife-edge test and its improved form, the phase contrast method. Mon. Not. R. Astron. Soc. **94**, 377–384 (1934)
77. Garbusi, E., Pruss, C., Osten, W.: Interferometer for precise and flexible asphere testing. Opt. Lett. **33**(24), 2973–2975 (2008)
78. Liesner, J., Garbusi, E., Pruss, C., Osten, W.: Verfahren und Messvorrichtung zur Vermessung einer optisch glatten Oberfläche, German Patent DE102006057606B4, 11 December 2008
79. Garbusi, E., Osten, W.: Perturbation methods in optics: application to the interferometric measurement of surfaces. J. Opt. Soc. Am. A. **26**, 2538–2549 (2009)
80. Baer, G., Schindler, J., Pruss, C., Siepmann, J., Osten, W.: Calibration of a non-null test interferometer for the measurement of aspheres and free-form surfaces. Opt. Express. **22**(25), 31200–31211 (2014)
81. Baer, G., Schindler, J., Siepmann, J., Pruss, C., Osten, W., Schulz, M.: Measurement of aspheres and free-form surfaces in a non-null test interferometer: reconstruction of high-frequency errors. Proc. SPIE. **8788**, 878818 (2013)
82. Baer, G., Garbusi, E., Lyda, W., Osten, W.: Automated surface positioning for a non-null test interferometer. Opt. Express. **49**(9), 095602 (2010)
83. Baer, G., Schindler, J., Pruss, C., Osten, W.: Correction of misalignment introduced aberration in non-null test measurements of free-form surfaces. J. Europ. Opt. Soc. **8**, 130874 (2013)
84. Malacara, D. (ed.): Optical Shop Testing, 3rd edn. Wiley, New York (2007)
85. Larkin, K.G.: Effective nonlinear algorithm for envelope detection in white light interferometry. J. Opt. Soc. Am. A. **13**, 832–843 (1996)
86. Sandoz, P., Devillers, R., Plata, A.: Unambiguous profilometry by fringe-order identification in white-light phase-shifting interferometry. J. Mod. Opt. **44**, 519–534 (1997)
87. Creath, K.: Calibration of numerical aperture effects in interferometric microscope objectives. Appl. Opt. **28**, 3333–3338 (1989)
88. Sheppard, C.J.R., Larkin, K.G.: Effect of numerical aperture on interference fringe spacing. Appl. Opt. **34**, 4731–4734 (1995)
89. Caber, P.J.: Interferometric profiler for rough surfaces. Appl. Opt. **32**, 3438–3441 (1993)
90. Itoh, M., Yamada, R., Tian, R., Tsai, M., Yatagai, T.: Broad-band light-wave correlation topography using wavelet transform. Opt. Rev. **2**(2), 135–138 (1995)
91. de Groot, P., Deck, L.: Surface profiling by analysis of White-light Interferograms in the spatial frequency domain. J. Mod. Opt. **42**, 389–401 (1995)
92. Sandoz, P.: Wavelet transform as a processing tool in White-light interferometry. Opt. Lett. **22**, 1065–1067 (1997)
93. Recknagel, R.-J., Notni, G.: Analysis of White light Interferograms. Opt. Comm. **148**, 122–128 (1998)
94. de Groot, P., de Lega, X.C., Kramer, J., Turzhitsky, M.: Determination of fringe order in white-light interference microscopy. App. Opt. **41**(22), 4517–4578 (2002)
95. de Groot, P., de Lega, X.C.: Signal modeling for low-coherence height-scanning interference microscopy. Appl. Opt. **43**(25), 4821–4830 (2004)
96. Larkin, K.G.: Efficient nonlinear algorithm for envelope detection in white light interferometry. J. Opt. Soc. Am. **A4**, 832–843 (1996)
97. Ai, C., Novak, E.: Centroid approach for estimation modulation peak in broad-bandwidth interferometry. U.S. Patent. **5**(633), 715 (1997)
98. Davidson, M., Kaufman, K., Mazor, I., Cohen, F.: An application of interference microscopy to integrated circuit inspection and metrology. Proc. SPIE. **775**, 233–240 (1987)
99. Kino, G.S., Chim, S.: Mirau Correlation Microscope. Appl. Opt. **29**, 3775–3783 (1990)
100. Lee, B.S., Strand, T.C.: Profilometry with a coherence scanning microscope. Appl. Opt. **29**, 3784–3788 (1990)

101. Dresel, T., Häusler, G.: Three dimensional sensing of rough surfaces by coherence radar. Appl. Opt. **31**, 919–925 (1992)
102. Häusler, G., Herrmann, J.M.: Physical limits of 3D-sensing. Proc. SPIE. **1822**, 150–158 (1992)
103. Häusler, G., Neumann, J.: Coherence radar-an accurate 3D sensor for rough surfaces. Proc. SPIE. **1822**, 200–205 (1992)
104. Mirau, A. H.: Interferometer, U. S. Patent US2612074 (A), 30 September 1952
105. Linnik, V.P.: Ein Apparat fur mikroskopisch-interferometrische Untersuchung reflektierender Objekte (Mikrointerferometer). Akad. Nauk S.S.S.R. Doklady. **21**(1), 18–23 (1933)
106. Machleidt, T., Jahn, R., Wenzel, K., Nestler, R., Franke, K.-H.: Materialeffekten auf der Spur, Laser þ Photonik no. 2/2014, 46–49 (2014)
107. Lewke, D., Schellenberger, M., Pfitzner, L., Fries, T., Tröger, B., Muehlig, A., Riedel, F., Bauer, S., Wihr, H.: Full Wafer Nanotopography Analysis on Rough Surfaces Using Stitched White Light Interferometry Images,. *ASMC 2013 SEMI Advanced Semiconductor Manufacturing Conference*, pp. 243–248. Saratoga Springs, NY (2013). https://doi.org/10.1109/ASMC.2013.6552812
108. Huntley, J.M., Saldner, H.O.: Temporal phase-unwrapping algorithm for automated fringe analysis. Appl. Opt. **32**, 3047–3052 (1993)
109. Huntley, J.M., Saldner, H.O.: Error-reduction methods for shape measurement by temporal phase unwrapping. J. Opt. Soc. Am. A. **14**, 3188–3196 (1997)
110. Saldner, H.O., Huntley, J.M.: Shape measurement by temporal phase unwrapping: comparison of unwrapping algorithms. Meas. Sci. Technol. **8**, 986–992 (1997)
111. Saldner, H.O., Huntley, J.M.: Temporal phase unwrapping: application to surface profiling of discontinuous objects. Appl. Opt. **36**, 2770–2775 (1997)
112. Huntley, J.M., Coggrave, C.R.: Progress in phase unwrapping. Proc. SPIE. **3407**, 86–93 (1998)
113. Paulson, L., Sjödahl, M., Kato, J., Yamaguchi, I.: Temporal phase unwrapping applied to wavelength-scanning interferometry. Appl. Opt. **39**, 3285–3288 (2000)
114. Hildebrand, B.P., Haines, K.A.: Multiple wavelength and multiple source holography applied to contouring generation. J. Opt. Soc. Am. **57**, 155–156 (1967)
115. Heflinger, L.O., Wuerker, R.F.: Holographic contouring via multifrequency lasers. Appl. Phys. Lett. **15**, 28–30 (1969)
116. Weigl, F.: Two-wavelength holographic interferometry for transparent media using a diffraction grating. Appl. Opt. **10**, 1083–1086 (1971)
117. Wyant, J.C.: Testing Aspherics using two-wavelength holography. Appl. Opt. **10**, 2113–2118 (1971)
118. Polhemus, C.: Two-wavelength interferometry. Appl. Opt. **12**, 2071–2074 (1973)
119. Petter, J.: Multi-wavelength interferometry for high precision distance measurement, Proc. OPTO 2009 & IRS^2 2009 (AMA-Science Service, Wunstorf), OP4, 129–132 (2009)
120. Abramson, N.: The Interferoscope: a new type of interferometer with variable fringe separation. Optik. **30**, 56–71 (1969)
121. Carisson, T.E., Abramson, N.H., Fischer, K.H.: Automatic measurement of surface height with the interferoscope. Opt. Eng. **35**, 2938–2942 (1996)
122. Schwider, J., Burow, R., Elssner, K.-E., Grzanna, J., Spolaczyk, R.: Semiconductor wafer and technical flat planeness testing interferometer. Appl. Opt. **25**, 1117–1121 (1986)
123. Boebel, D., Packross, B., Tiziani, H.J.: Phase shifting in an oblique incidence interferometer. Opt. Eng. **30**(12), 30–35 (1991)
124. Spür, G., Nyarsik, L., Körner, K.: Imaging characteristics of prism interferometers. Proc. SPIE. **1983**, 702–703 (1993)
125. Birch, K.G.: Application of the "Interferoscope" to spherical and aspherical surfaces. Optik. **36** (4), 399–409 (1972)
126. Birch, K.G.: Oblique incidence interferometry applied to non-optical surfaces. J. Phys. E. **6**, 1045–1048 (1973)

127. Hariharan, P.: Improved oblique-incidence interferometer. Opt. Eng. **14**, 257–258 (1974)
128. Järisch, W., Makosch, G.: Interferometric surface mapping with variable sensitivity. Appl. Opt. **17**, 740–742 (1978)
129. Dresel, T., Schwider, J., Wehrhahn, A., Babin, S.: Grazing incidence interferometry applied to the measurement of cylindrical surfaces. Opt. Eng. **34**, 3531–3535 (1995)
130. Kulawiec, A.W., Fleig, J.F., Bruning, J.H.: Interferometric measurements of absolute dimensions of cylindrical surfaces, 1997 Annual Meeting of the ASPE, Norfolk, VA, October 5–10, (1997)
131. Hizuka, M.: Oblique incidence interferometer with fringe scan drive. U.S. patent. **5**(786), 896 (1998)
132. de Groot, P.: Diffractive grazing-incidence interferometer. Appl. Opt. **39**(10), 1527–1530 (2000)
133. Olsson, A., Tang, C.L.: Dynamic interferometry techniques for optical path length measurements. Appl. Opt. **20**, 3503–3507 (1981)
134. Kikuta, H., Iwata, K., Nagata, R.: Distance measurement by the wavelength shift of laser diode light. Appl. Opt. **25**, 2976–2980 (1986)
135. Sasaki, O., Yoshida, T., Suzuki, T.: Double sinusoidal phase modulating laser diode interferometer for distance measurement. Appl. Opt. **30**, 3617–3621 (1991)
136. Beheim, G., Fritsch, K.: Remote displacement measurements using a laser diode. Electron. Lett. **21**, 93–94 (1985)
137. den Boef, A.J.: Interferometric laser range finder using a frequency modulated diode laser. Appl. Opt. **26**, 4545–4550 (1987)
138. Suematsu, M., Takeda, M.: Wavelength-shift interferometry for distance measurements using the Fourier transform technique for fringe analysis. Appl. Opt. **30**, 4046–4055 (1991)
139. Gabor, D.: Theory of communication. J. Inst. Electr. Eng. **93**, 429–441 (1946)
140. Gabor, D.: A new microscopic principle. Nature. **161**, 777–778 (1948)
141. Gabor, D.: Microscopy by reconstructed wavefronts. Proc. Royal Soc. **A197**, 454–487 (1949)
142. Gabor, D.: Microscopy by reconstructed wavefronts: II. Proc. Phys. Soc. **B64**, 449–469 (1951)
143. Goodman, J.W., Lawrence, R.W.: Digital image formation from electronically detected holograms. Appl. Phys. Lett. **11**, 77–79 (1967)
144. Pluta, M.: Advanced Light Microscopy, vol. 2, pp. 282–352. Elsevier, Amsterdam (1989), Chap. 11)
145. Hariharan, P.: Optical Holography. Cambridge University, Cambridge (1996)
146. Kreis, T.: Handbook of Holographic Interferometry: Optical and Digital Methods. Wiley VCH, Weinheim (2005)
147. Schnars, U., Jüptner, W.: Digital Holography: Digital Hologram Recording, Numerical Reconstruction, and Related Techniques. Springer, Berlin (2005)
148. Kim, M.K.: Principles and techniques of digital holographic microscopy. SPIE Rev. **1**(1), 018005 (2010). https://doi.org/10.1117/6.0000006
149. W. Osten, P. Ferraro, Digital holography and its application in MEMS/MOEMS inspection, in Optical Inspection of Microsystems, ed. by W. Osten, 1 (CRC Press/Taylor and Francis Group, Boca Raton, 2006), 351–426
150. Cuche, E., Bevilacqua, F., Depeursinge, C.: Digital holography for quantitative phase-contrast imaging. Opt. Lett. **24**, 291–293 (1999)
151. Cuche, E., Marquet, P., Depeursinge, C.: Simultaneous amplitude contrast and quantitative phase-contrast microscopy by numerical reconstruction of Fresnel off-axis holograms. Appl. Opt. **38**, 6994–7001 (1999)
152. Schnars, U., Jüptner, W.: Digital recording and numerical reconstruction of holograms. Meas. Sci. Technol. **13**, R85–R101 (2002)
153. Kim, M.K., Yu, L.F., Mann, C.J.: Digital holography and multi-wavelength interference techniques. In: Poon, T.-C. (ed.) Digital holography and three dimensional display: principles and applications, pp. 51–72. Springer, Boston (2006)

154. Kreis, T., Jüptner, W.: Suppression of the dc term in digital holography. Opt. Eng. **36**, 2357–2360 (1997)
155. Cuche, E., Marquet, P., Depeursinge, C.: Spatial filtering for zero-order and twin-image elimination in digital off-axis holography. Appl. Opt. **39**, 4070–4075 (2000)
156. Sirat, G., Psaltis, D.: Conoscopic holography. Opt. Lett. **10**(1), 4–6 (1985)
157. Sirat, G., Psaltis, D.: Conoscopic holograms. Opt. Comm. **65**(4), 243–249 (1988)
158. Sirat, G.: Conoscopic holography I. basic principles and physical basis. J. Opt. Soc. Am. **A9** (1), 70–83 (1992)
159. Sirat, G.: Conoscopic holography II. Rigorous derivation. J. Opt. Soc. Am. **A9**(1), 84–90 (1992)
160. Hartmann, J.: Bemerkungen über den Bau und die Justierung von Spektrographen Z. Instrumentenkunde. **20**, 17-27–47-58 (1900)
161. Shack, R.B., Platt, B.C.: Production and use of a lenticular Hartmann screen. J. Opt. Soc. Am. **61**, 656–660 (1971)
162. Platt, B.C., Shack, R.B.: History and principle of Shack-Hartmann Wavefront sensing. J. Refract. Surg. **17**, 573–577 (2001)
163. Pfund, J., Lindlein, N., Schwider, J.: Dynamic range expansion of a Shack-Hartmann sensor by use of a modified unwrapping algorithm. Opt. Lett. **23**, 995–997 (1998)
164. Olivier, S., Laude, V., Huignard, J.-P.: Liquid crystal Hartmann wave-front scanner. Appl. Opt. **39**, 3838–3846 (2000)
165. Liesener, J., Seifert, L., Tiziani, H.J., Osten, W.: Active wavefront sensing and wavefront control with SLMs. Proc. SPIE. **5532**, 147–158 (2004)
166. Stuerwald, S., Schmitt, R.: DMD-based scanning of steep wavefronts for optical testing of freeform optics. Proc. SPIE. **8618**, 8618–8619 (2013)
167. Li, W., Bothe, T., Koplow, C., Jüptner, W.: Evaluation methods for gradient measurement techniques. Proc. SPIE. **5457**, 300–311 (2004)
168. Bothe, T., Li, W., von Kopylow, C., Jüptner, W.: High-resolution 3D shape measurement on specular surfaces by fringe reflection. Proc. SPIE. **5457**, 411–422 (2004)
169. Bothe, T., Li, W., Kopylow, C., Jüptner, W.: In: Osten, W. (ed.) Fringe 2005 - 5th International Workshop on Automatic Processing of Fringe Patterns, pp. 362–371. Springer, Berlin, Heidelberg (2005)
170. Häusler, G.: Verfahren und Vorrichtung zur Ermittlung der Form oder der Abbildungseigenschaften von spiegelnden oder transparenten Objekten, German Patent DE 19944354 A1, 12 April 2001
171. Knauer, M.C., Häusler, G.: R. Lampalzer, Verfahren und Vorrichtung zur dreidimensionalen Vermessung der Form und der lokalen Oberflächennormalen von vorzugsweise spiegelnden Objekten, German patent DE 102004020419 (2004)
172. Knauer, M.C., Kaminski, J., Häusler, G.: Phase measuring Deflectometry: a new approach to measure specular free-form surfaces. Proc. SPIE. **5457**, 366–376 (2004)
173. Häusler, G., Knauer, M.C., Faber, C., Richter, C., Peterhänsel, S., Kranitzky, C., Veit, K.: Deflectometry: 3D-metrology from nanometer to meter. In: Osten, b.W., Kujawinska, M. (eds.) Fringe 2009: the 6th International Workshop on Advanced Optical Metrology, pp. 416–421. Springer, Berlin (2009)
174. Kugimiya, K.: Characterization of polished mirror surfaces by the "Makyoh" principle. Mater. Lett. **7**(5–6), 229–233 (1998)
175. Tokura, S., Fujino, N., Ninomiya, M., Masuda, K.: Characterization of mirror-polished silicon wafers by Makyoh method. J. Crystal Growth. **103**(1–4), 437–442 (1990)
176. Riesz, F.: Geometrical optical model of the image formation in Makyoh (magic mirror) topography. J. Phys. D. **33**(19), 3033–3040 (2000)
177. Vorburger, T.V., Marx, E., Lettieri, T.R.: Regimes of surface roughness measurable with light scattering. Appl. Opt. **32**(19), 3401–3408 (1993)
178. ASTM Standard F1048–87: Standard Test Method for Measuring the Effective Surface Roughness of Optical Components by Total Integrated Scattering (1987) (Reapproved 1999)

179. SEMI MF 1048-1109: Test method for measuring the effective surface roughness of optical components by total integrated scattering. Semiconductor Equipment and Materials International (2009)
180. Bennett, H.E., Porteus, J.O.: Relation between surface roughness and specular reflection at Normal incidence. J. Opt. Soc. Am. **51**, 123–129 (1961)
181. ASTM Standard E2387–05: Standard practice for Goniometric Optical Scatter Measurements (2011)
182. SEMI ME 1392-1109: Guide for angle resolved optical scatter measurements on specular or diffuse surfaces. Semiconductor Equipment and Materials International (2009)
183. Elson, J.M., Benett, J.M.: Calculation of the power spectral density from surface profile data. Appl. Opt. **34**(1), 201–208 (1995)
184. Brodmann, R., Gast, T., Thurn, G.: An optical instrument for measuring the surface roughness in production control. Ann. CIRP. **33**(/1), 403–406 (1984)
185. Brodmann, R., Gerstorfer, O., Thurn, G.: Optical roughness measuring instrument for fine-machined surfaces. Opt. Eng. **24**, 410–413 (1985)
186. Goodman, J.W.: Statistical Properties of Laser Speckle Patterns, in Laser speckle and related topics, Topics in Physics, vol. 9, pp. 9–75. Springer, New York (1975)
187. Parry, G.: Speckle Patterns in Partially Coherent Light, in Laser speckle and related topics, Topics in Physics, vol. 9, pp. 77–121. Springer, New York (1975)
188. Fleischer, J., Ruffing, B.: Spectral correlation of partially or fully developed speckle patterns generated by rough surfaces. J. Opt. Soc. Am. **A2**, 1637–1643 (1985)
189. Goodman, J.W.: Statistical Optics. Wiley, New York (1985)
190. Sirohi, R.S.: Speckle Metrology. Dekker, New York (1993)
191. Patzelt, S.: Simulation und experimentelle Erprobung parametrisch-optischer Rauheitsmessprozesse auf der Basis von kohärentem Streulicht und Speckle-Korrelationsverfahren, Ph.D. thesis (Verlagshaus Mainz GmbH, Aachen, 2010)
192. Quinten, M.: Practical Determination of Optical Constants from Spectral Measurements. BoD-Book on Demands, Norderstedt (2018)
193. Quinten, M.: A Practical Guide to Optical Metrology for Thin Films. Wiley-VCH, Weinheim (2012)
194. Humlicek, J., Carriga, M., Alonso, M.I., Cardona, M.: Optical spectra of $Si_xGe_{(1-x)}$ alloys. J. Appl. Phys. **65**, 2827–2832 (1989)
195. Palik, E.D. (ed.): Handbook of Optical Constants of Solids I. Academic, San Diego (1985)
196. Gerfin, T., Grätzel, M.: Optical properties of tin-doped indium oxide determined by spectroscopic ellipsometry. J. Appl. Phys. **79**, 1722–1729 (1996)
197. Levenberg, K.: A method for the solution of certain problems in least squares. Q. Appl. Math. **2**, 164–168 (1944)
198. Marquardt, D.: An algorithm for least-squares estimation of nonlinear parameters. SIAM J. Appl. Math. **11**, 431–441 (1963)
199. Nelder, J.A., Mead, R.: A simplex method for function minimization. Comp. J. **7**, 308–313 (1965)
200. Press, W.H., Teukolsky, S.A., Vetterling, W.T., Flannery, B.P.: Numerical Recipes – the Art of Scientific Computing, 3rd edn. Cambridge University Press, Cambridge (2007)
201. Garnett, J.C.M.: Colours in metal glasses and in metallic films. Phil. Trans. Royal Soc. London A. **203**, 385–420 (1904)
202. Bruggeman, D.A.G.: Berechnung verschiedener physikalischer Konstanten von heterogenen Substanzen. I. Dielektrizitätskonstanten und Leitfähigkeiten der Mischkörper aus isotropen Substanzen. Ann. Phys. (Leipzig). **24**, 636–679 (1935)
203. Drude, P.: Ueber Oberflaechenschichten, I. Theil, Ann. Physik u. Chemie 36, 532–560, and II. Theil, 865–897 (1889)
204. Rothen, A.: The Ellipsometer, an apparatus to measure thickness of surface films. Rev. Sci. Instrum. **16**(2), 26–30 (1945)

205. Azzam, R.M.A., Bashara, N.M.: Ellipsometry and Polarized Light, 2nd edn. Elsevier, Amsterdam (1987)
206. Tompkins, H.G.: A User's Guide to Ellipsometry. Dover Publications Inc., New York (2006)
207. Tompkins, H.G., McGahan, W.A.: Spectroscopic Ellipsometry and Reflectometry. Wiley, New York (1999)
208. Tompkins, H.G., Irene, E.A., Haber, E.A.: Handbook of Ellipsometry (Materials Science and Process Technology). William Andrew Inc., New York (2005)
209. Tompkins, H.G., Irene, E.A. (eds.): Handbook of Ellipsometry. Springer, Berlin (2006)
210. Fujiwara, H.: Spectroscopic Ellipsometry: Principles and Applications, 1st edn. Wiley, New York (2007)
211. Röseler, A.: Infrared Spectroscopic Ellipsometry. Akademie-Verlag, Berlin (1990)

Chapter 6
Imaging Methods

Abstract The methods presented before in Chaps. 2, 3, and 5 concentrate on form deviations, topography, roughness, and waviness. In principle, also dimensional deviations, position deviations and defect inspection can be accomplished by these methods. More common is to use imaging and image processing as the main processes for measurement and evaluation for these more coarse deviations.

This chapter comprises the following imaging methods:

- Industrial Image Processing,
- Shape from Shading,
- Hyperspectral Imaging,
- Scanning Electron Microscopy,
- Optical Coherence Tomography, and
- Terahertz Spectroscopy.

6.1 Industrial Image Processing

In manufacturing the *industrial image processing* (IIP) or *machine vision* is a key technology with high dynamics. Anyhow, despite the enormous progress in different fields, mainly the computer technology, there does not exist an all-in-one device suitable for every purpose but the device must be adjusted individually to the corresponding application to have the most efficient system for this application. IIP is a relatively young technology starting from 1975.

Although IIP bases upon digitally recorded images the output of a IIP system is not further an image but can have another semantic quality. The information contained in the output is therefore not further determined for the human eye but is used in various automation processes (control, check, transformation, etc.).

In manufacturing it is used

- To identify, to sort, and to position workpieces and to check the quality of workpieces,
- To characterize defects of faulty items, and

M. Quinten, *A Practical Guide to Surface Metrology*, Springer Series in Measurement Science and Technology,
https://doi.org/10.1007/978-3-030-29454-0_6

- To monitor, to control, and to optimize processes.

These activities include, among others, electronics component manufacturing, quality textile production, metal product finishing, glass manufacturing, machine parts, printing products, semiconductor inspection, optical character recognition, layer thickness determination, and surface inspection.

In most cases industrial automation systems are designed to inspect only known objects at fixed positions. The design and development of a successful machine vision system vary depending on the application domain and are related to the tasks to be accomplished (environment, speed, etc.). The contactless measurement of workpieces with IIP has the big advantage that the measurement can be carried out very fast within the production cycle time. Integrated in the production process IIP measures each single workpiece without decreasing the production cycle time. Industrial Image Processing stands for acceptable accuracy, high robustness, high reliability, and high mechanical and thermal stability.

Industrial image processing (IIP) is based upon taking photographs of the workpieces or videos and evaluating them with digital image processing. The most important components of an industrial image processing system are

- Illumination
- Lens System
- Camera
- Image Processing Hardware
- Digital Image Processing Software,
- Interfaces, and
- Mechanics.

Figure 6.1 depicts the schematic setup of an IIP system. In the following the respective components are discussed.

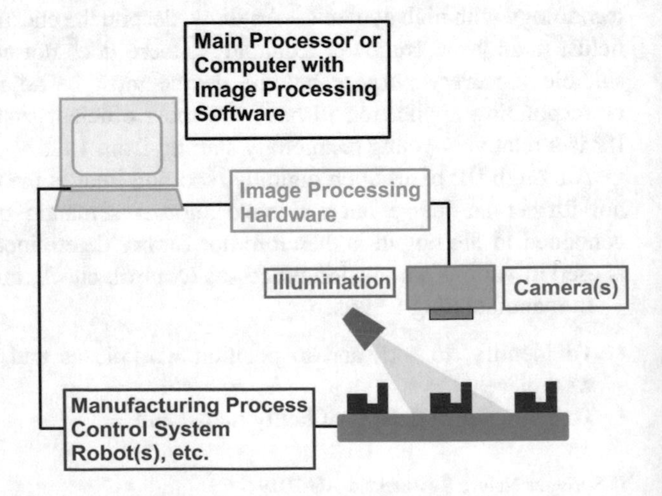

Fig. 6.1 Schematic setup of an image processing system

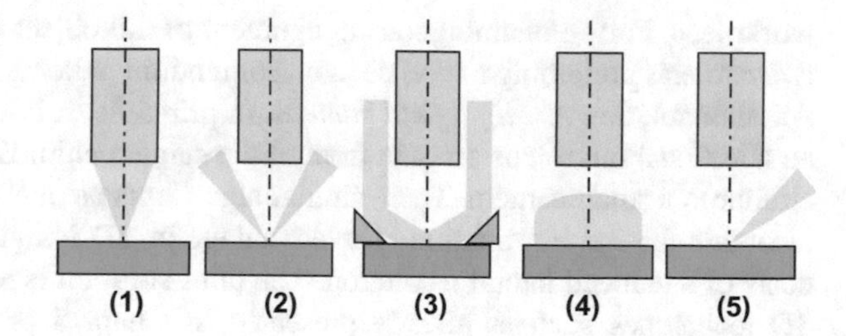

Fig. 6.2 Individual types of illumination: (*1*) coaxial illumination, (*2*) ring light, (*3*) dark field, (*4*) diffuse illumination, (*5*) point light source

6.1.1 Illumination

The illumination is an essential component of IIP as the features which are later evaluated by the inspection software are basically affected. The light source needs to be adjusted so that the illumination is appropriate for the sample material and geometry. Stable results require constant sufficiently intensive lighting conditions that leads to a sufficient contrast between parts of the workpiece. Secondary light sources should be excluded because they can have a strong impact on the inspection. With proper filters or light gates the influence of daylight and secondary light sources can be reduced. It is a special ability to achieve a good illumination level. Figure 6.2 shows individual types of possible illuminations.

6.1.2 Camera

Images are usually acquired by one or more cameras placed at the scene under inspection. The positions of the cameras are usually fixed. Cameras contain the camera chip (CMOS or CCD) with $N \times M$ pixels of certain size, the camera electronics which provides digital processing of the camera data, and a lens system that reproduces the illuminated workpiece onto the camera. The lenses determine the size of the imaged section of the workpiece, the field of view. Highly precise measuring tasks require telecentric lenses and, if possible, also telecentric lighting to avoid distortions and complex angle corrections in the test routine. Sometimes special lenses may be reasonable and even necessary.

The used cameras are also able to record moving objects without blurring if a short exposure time is chosen. Matrix cameras are easier to handle than line cameras. Matrix cameras deliver images which can be examined immediately on the monitor. Line cameras are more suitable for the inspection of lengths. The requirements on a camera for IIP are related to signal-to-noise ratio, frame rate, sensitivity, dynamic range, light exposure, and accuracy.

The term "camera" implies the projection of a 3D scene onto a 2D image. Many tasks in IIP can however favorably be solved with 3D techniques. One solution is to use two cameras for stereoscopy. The two images get evaluated with geometrical triangulation to obtain a 3D image. This technique even allows movement of the workpiece. For an unambiguous assignment of the object point to the pixels of the two cameras preferably reference marks or random patterns are present on the object. Another solution is a *light field camera*. In principle, it is a normal camera but with an additional microlens array in front of the camera chip. Each lens of the array acts similar to a small camera. Then, similar algorithms as in stereoscopy can be applied to obtain the 3D information. The advantage of 3D imaging is yet obtained at the costs of a reduced lateral resolution. The third solution is to combine a camera with 3D techniques such as already presented in Chap. 5 or to use these techniques directly. These techniques comprise light sectional methods, digital holographic microscopy, white light interferometry, focus variation, and deflectometry. In addition, optical coherence tomography (see Sect. 6.5) and methods not treated in this book like photogrammetry, LIDAR (light detection and ranging), X-ray tomography, and time-of-flight methods are in use for this purpose. In time-of flight methods the time that a light pulse needs for the way to the object and back is measured for each point of the object. The needed time is proportional to the distance whereby a height distribution of the surface of the object is obtained. Recently, a new technique called *Inline Computational Imaging* (ICI) has been developed [1–3] that combines light field techniques and stereoscopy techniques. It is based on a single camera with non-telecentric objective where every two pixel rows are used as line scan camera. When the object moves under the camera a stack of images is recorded. Each image has a slightly different illumination condition (light field) and a slightly different angle of view (stereoscopy).

6.1.3 Image Processing Hardware

Image processing hardware calculates measurement categories, controls signals for sorters, or quality parameters for processes and products. Intelligent cameras are already equipped with built-in evaluation electronics.

When the process is highly time-constrained or computationally intensive and exceeds the processing capabilities of the main processor, application specific hardware (e.g. DSPs, ASICs, or FPGAs) is employed to alleviate the problem of processing speed.

Also the PC may be equipped with PCI cards which are able to take over digitalization as well as more complex functions for image preprocessing or evaluation. In the basic case this PCI card merely forms the interface to the camera and digitalizes the analog camera signals.

6.1.4 Digital Image Processing Software

The core competence of many vendors lies in the field of software development. The evaluation software must be structured so that the images are evaluated with respect to the particular task. The software also passes the results to the control system, for example it sends a signal to the mechanical equipment to sort out defective pieces. For this highly sophisticated software with artificial intelligence, neuronal networks, and deep learning is used.

6.1.5 Mechanics

The fixation of camera and illumination, the moving of the workpiece, and the product handling are considered in the mechanics. The quality of the mechanics has a strong influence on how the complete system functions. It is important that fixations are stable and that the workpiece is well lead. If pieces are not positioned properly in front of the camera the evaluation process becomes more difficult and possibly slows down the production process.

The automated inspection of surfaces is one of the most demanding tasks of the IIP. The challenge is to recognize 3D defects on colored or textured surfaces. The reason is that a 2D image does not yield unambiguously the information on whether a 3D defect is recognized or only a spot on the surface. A human observer would turn the workpiece in the light to improve the contrast. In automated inspection systems this can be mimicked by a series of test consoles with different illumination conditions. Then, multiple photographs are taken and correspondingly evaluated.

Before an IIP system gets selected and installed the following questions should be answered:

- How many inspections per second are requested?
- How fast do the parts move?
- How large is the inspection volume?
- Which accuracy is requested?
- Does one needs more than one view?
- Is color recognition necessary?
- Which actions follow based upon the results of the data evaluation?
- Which communication with the facility shall be enabled?

The answers to these questions help to select the IIP system that is best equipped and adjusted to the customer's needs.

6.2 Shape from Shading

Shape from Shading (SfS) is a method to retrieve 3D information on the shape of a body from the shading when this body is illuminated with parallel light. For this a camera takes a picture from the body which gets illuminated under a fixed direction. This is illustrated in Fig. 6.3.

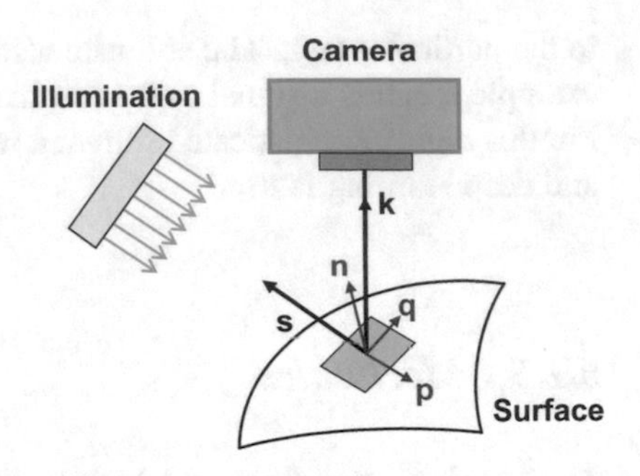

Fig. 6.3 Sketch of the principle of shape from shading

Due to the shading the recorded light intensity has gradients in x- and y-direction of a reference coordinate system. The measured intensity $I(x, y)$ at a pixel (m, n) depends upon the direction of the illumination, the position of the camera and its sensitivity, the local reflectivity of the illuminated surface, and the angle enclosed between the known direction **s** of the illumination and the unknown normal vector **n** in the point (x, y). As the normal vector can be retrieved from the gradients in x- and y-direction, $p = \partial z/\partial x$ and $q = \partial z/\partial y$ as

$$\mathrm{n} = \frac{1}{\sqrt{1 + p^2 + q^2}} \begin{pmatrix} -p \\ -q \\ 1 \end{pmatrix} \tag{6.1}$$

and these gradients are identical to the gradients in the recorded intensity distribution, a conditional equation

$$\mathrm{I}(x, y) = \rho \cdot \mathrm{n} \cdot \mathrm{s} = \rho \cdot \left(s_z + p \cdot s_x + q \cdot s_y\right) \tag{6.2}$$

can be set for the three unknown parameters ρ, p, and q and the recorded intensity at position (x, y). This single equation is not sufficient. At least, three equations must be used for an unambiguous determination of the three unknown parameter. This is established by two additional pictures for two further illuminations under different but well-known illumination directions. In practice, even four pictures for four different illumination directions are used. Image processing finally delivers information on tilt and curvature from which the surface of the workpiece is reconstructed.

This method allows for recognition of form deviations like cracks, scratches, pores, or grooves, particular on plane surfaces. An example is given in Fig. 6.4 where four pictures of a coin were taken under different illumination directions and then evaluated for curvature, gradients in x- and y-direction, and texture. The pictures are courtesy of SAC Sirius Advanced Cybernetics GmbH, Karlsruhe, Germany, recorded with the trevista® system.

It was Horn [4, 5] who formulated first the SfS problem simply and rigorously as that of finding the solution of a nonlinear first-order partial differential equation

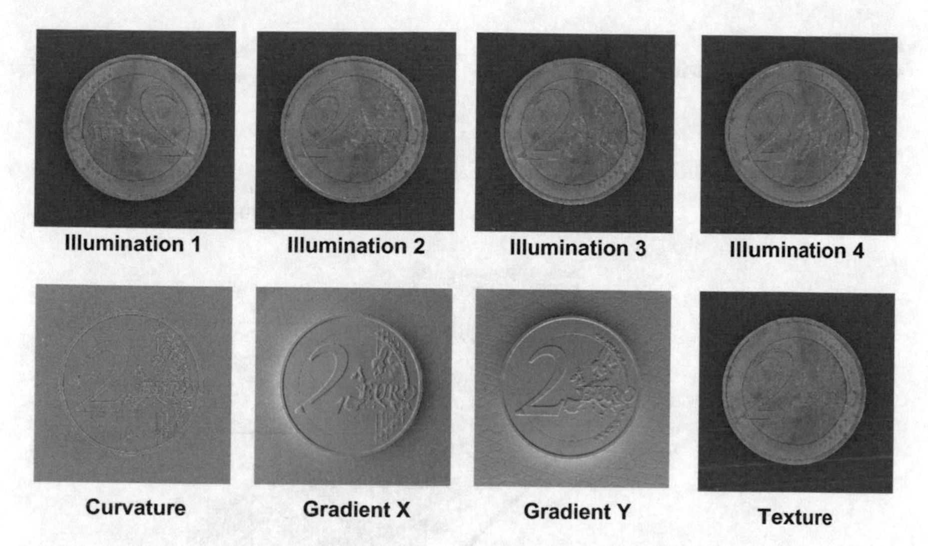

Fig. 6.4 Defect recognition by shape from shading. (Courtesy of SAC Sirius Advanced Cybernetics GmbH, Karlsruhe, Germany)

called the brightness equation. The used algorithm of today [6, 7] is mathematically elaborated and needs three Fourier transforms. Today, the SfS problem is known to be an ill-posed problem. A number of articles show that the solution is not unique and particularly encounters concave/convex ambiguities. Being aware of the difficulties one has to claim that all the parameters of the light source, the surface reflectance, and the camera are well-known and the light intensity is constant. Moreover, the object should be a Lambertian radiator with known albedo and should have a continuous surface without steep edges.

6.3 Hyperspectral Imaging

Classical imaging for industrial image processing and spectroscopy have independently proven to be valuable methods for a wide range of applications. Images commonly contain no or only little spectral information while spectroscopy is severely limited when inhomogeneous samples are to be measured. *Hyperspectral Imaging* (HSI) combines both technologies and allows thereby qualitative and quantitative studies of both, spatial and spectral features. The purpose of HSI is to produce a picture of the target with each pixel containing an entire spectrum. This spectral oversampling allows to distinguish differences between material classes.

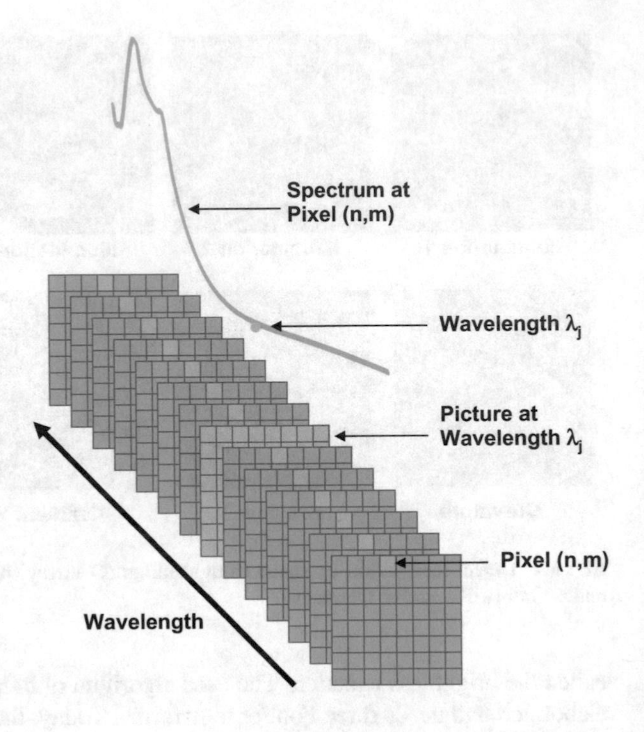

Fig. 6.5 Principle of spectral scanning HSI with hyperspectral cube

There are two common methods to generate hyperspectral images:

- Spectral (wavelength) scanning HSI or staring imagers and
- spatial scanning HSI or push-broom imagers.

Which method is more comfortable depends upon the application. In combination with advanced and fast data evaluation algorithms it is nowadays possible to analyze a wide range of samples with high spectral and spatial resolution.

Spectral Scanning HSI (Staring Imagers)
In *spectral scanning HSI* each camera output represents a single-colored image of the investigated scene. HSI devices for spectral scanning typically use optical band-pass filters to change the wavelength from image to image. Then, one obtains a series of images for which when tracing each pixel (n, m) a complete spectrum is contained in the third dimension of the so-called hyperspectral cube (see Fig. 6.5).

Spatial Scanning HSI (Pushbroom Imagers)
In *spatial scanning HSI* the camera images the scene line by line using the so-called pushbroom scanning mode. One narrow spatial line in the scene is imaged at a time onto a slit, e.g. N points x_n in x-direction at scanning position y_m. The slit image gets split into its spectral components. Then, on the camera one direction is used for the spectrum at point (x_n, y_m) and the second direction is used

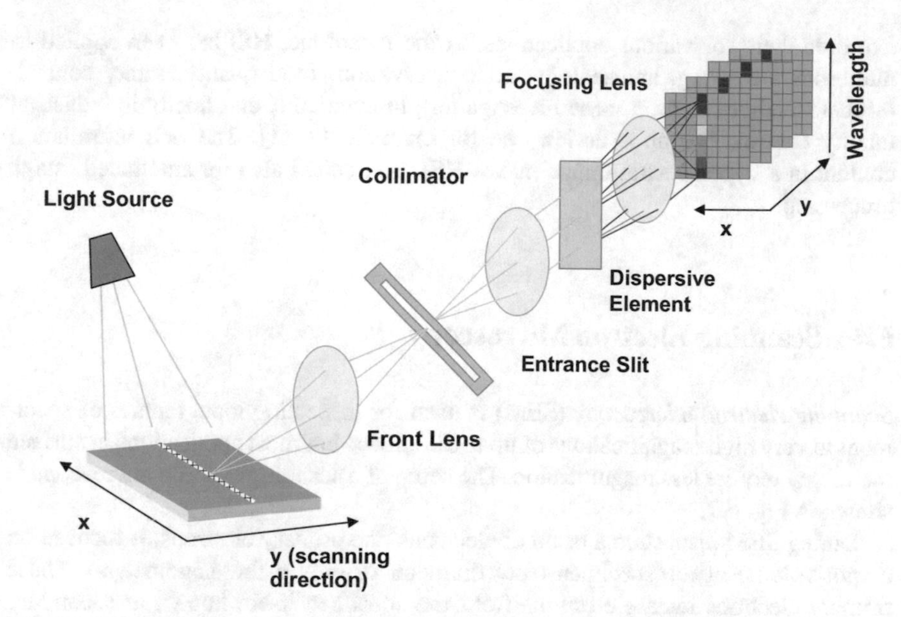

Fig. 6.6 Principle of spatial (pushbroom) scanning HSI with hyperspectral cube

for imaging the N points x_n. Afterwards, the sample is moved to the next scanning position y_{m+1}. The principle of the spatial scanning HSI is illustrated in Fig. 6.6.

In any case the data in the hyperspectral cube contain a colored image of the sample with additional information on the spectral reflectance or radiance at each pixel (n, m) or the corresponding position (x_n, y_m). So, hyperspectral images provide much more detailed information about the sample than a normal color camera which only acquires three different spectral channels corresponding to the visual primary colors red, green, and blue. Hyperspectral imaging allows to classify the objects in the scene based on their spectral properties.

Due to the additional recording of a spectrum, information on color and fluorescence are available if visible light is used. Information on chemical composition and functional groups are available using the near infrared spectral region.

The use of a microscope in a HSI system allows to examine structures of a few hundred nanometers (in the visible spectral range) or a few microns (in the near infrared range) as well as thickness determination of thin layers with methods described before in Sect. 5.10.

A particular form of HSI is realized in the *Confocal Raman Imaging*. Here, the sample is scanned in one focus position of a microscope point-by-point and line-by-line and at every image pixel a complete Raman spectrum is taken. Then, the spectra are analyzed to get the distribution of chemical sample properties in the focal plane. Taking a stack of images by varying the focus position, the geometry of the sample can be reconstructed in 3D.

Hyperspectral imaging has a long tradition over four decades. It originates from satellite based/airborne remote sensing and military target detection and has been explored there for various applications. In the meantime, HSI has been applied to numerous areas, e.g. archaeology, art conservation, food quality, safety control, forensic medicine, crime scene investigation, biomedicine, and finally in industrial quality control. For an overview see for example [8–11]. The rich information content in a hyperspectral image makes HSI well suited also for automated image processing.

6.4 Scanning Electron Microscopy

Scanning electron microscopy (SEM) is used for inspecting topographies of specimens at very high magnifications of up to 500,000× but most applications in surface metrology require less magnification. The setup of a scanning electron microscope is shown in Fig. 6.7.

During SEM inspection a beam of electrons, the primary electrons, is focused on a spot volume of the specimen (spot diameter of only a few nanometers). These primary electrons release electrons from the specimen itself, known as secondary electrons. They are attracted and collected by a positively biased detector and then translated into a signal. The yield of secondary electrons depends upon the angle of incidence, the place where the primary electrons hit the specimen, the material of the specimen, the crystalline structure of the material, and electrical potentials. To generate the SEM image the electron beam is swept across the specimen area, producing many such signals. The signals get amplified, analyzed, and translated into images of the topography of the specimen. Finally, the image is shown on a display. The image resembles that seen through an optical lens but at a much higher

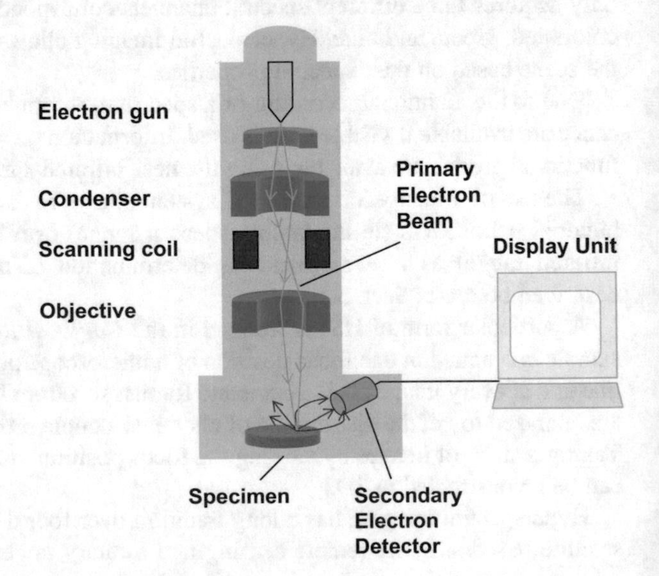

Fig. 6.7 Principle of a Scanning Electron Microscope (SEM)

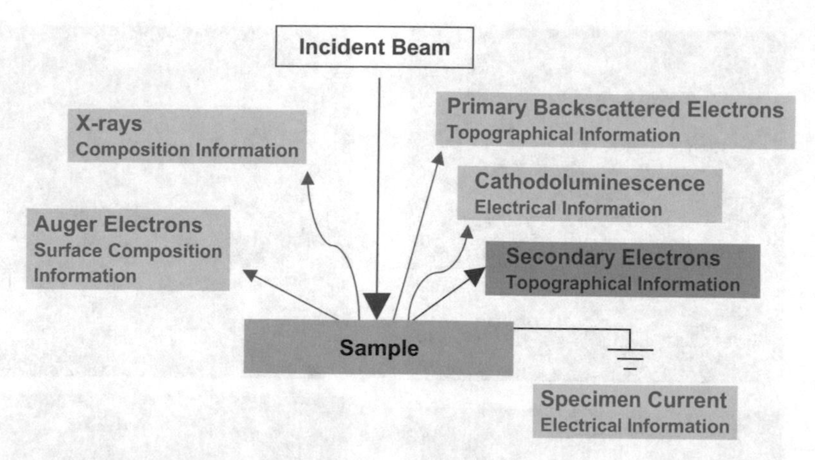

Fig. 6.8 Interaction of the electron beam with the target

resolution. SEMs typically measure surface topography on a much smaller spatial wavelength scale compared to a stylus tip and optical sensors. This yields extraordinary 3D impressions and the lateral resolution is better than 5 nm, however the depth of field is pretty large so that no true depth or height information is obtained.

When the primary electrons hit the specimen not only secondary electrons are generated. The interaction of the electrons with the material of the sample yields in addition backscattered (reflected) electrons, Auger electrons, characteristic X-rays, and photons (see Fig. 6.8).

Backscattered electrons can be used to distinguish one material from another. For this they are collected and also displayed in an image. Backscatter imaging can distinguish elements with atomic number differences of at least 3. Backscattered electron images provide information about the distribution but not the identity of different elements in the sample.

The energy or wavelength of the characteristic X-rays can be measured by *Energy-Dispersive X-Ray spectroscopy* (EDX) and can be used to identify and measure the amount of elements in the sample. Also a map of the distribution of the elements can be made.

Already in 1935 Max Knoll [12] produced a photograph with an electron beam scanner. A true scanning electron microscope with high magnification was invented by Manfred von Ardenne in 1937 [13]. Ardenne applied the scanning principle not only to achieve magnification but also to eliminate the chromatic aberration which is otherwise inherent in the electron microscope. He further discussed the various detection modes, possibilities and theory of SEM [14], together with the construction of the first high magnification SEM [15]. For an overview on SEM the reader is referred to the monographs of L. Reimer [16] and J. Goldstein et al. [17]. SEM stereoscopy and angle-resolved SEM are listed as techniques in ISO 25178-6.

The main drawback with SEM is that it is essentially a two-dimensional technique due to large depth of field. Real 3D information can be obtained from

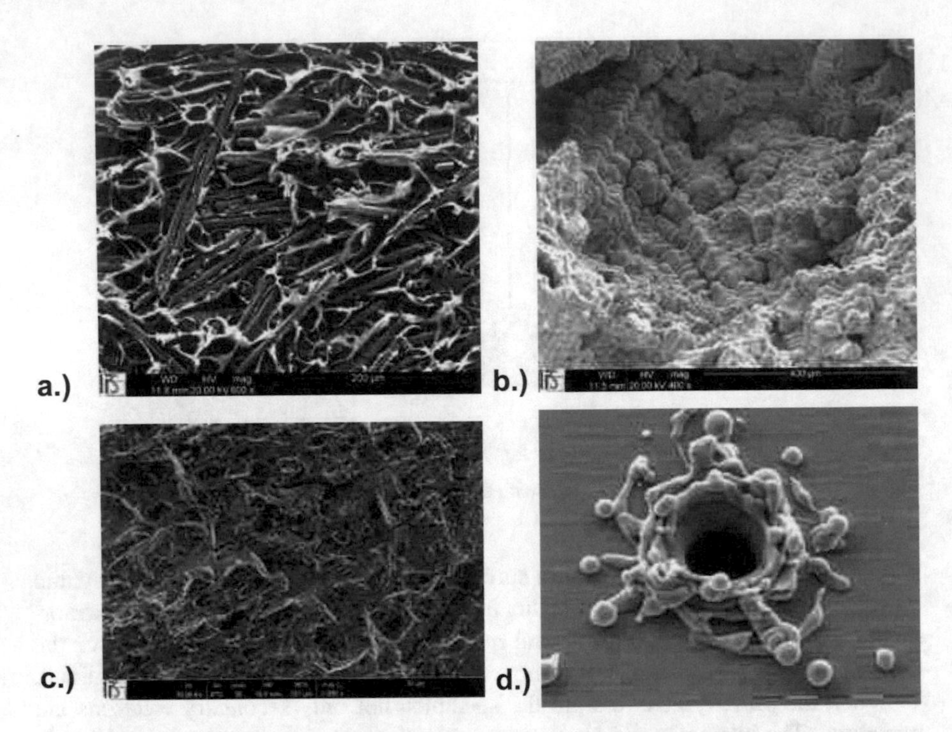

Fig. 6.9 Exemplaric SEM pictures: (**a**) ductile forced fracture in a glass-fiber reinforced plastics, (**b**) shrink hole (bubble) in red bronze, (**c**) stress-corrosion cracking in brass, (**d**) laser burned hole in steel. The pictures in (**a, c**) are courtesy of the Institut für Schadenverhütung und Schadenforschung der öffentlichen Versicherer e.V. (www.ifs-ev.org), Kiel, Germany, and in (**d**) is courtesy of the Laserzentrum der Fachhochschule Münster, Steinfurt, Germany

some surfaces by tilting the sample and using a stereo imaging or angle-resolved scanning techniques [17].

SEM is a laborious technique that requires vacuum and has certain requirements on the samples. Samples must be conductive, otherwise they must be coated with a thin film of a conducting material (gold, chromium, carbon) to prevent electrostatic charging. Samples need not to outgas. Biological samples must be fixed and freeze-dried to retain integrity under high vacuum conditions. Despite these disadvantages SEM has a wide field of applications. The main application is failure analysis (cracks, fissures, corrosion, contaminations, material defects, faults in manufacturing and processing), followed by materials science (microstructure, porosity, chemical composition, surface analysis, changes caused by temperature), quality control, and forensics. Often, SEM is used in combination with EDX. In the following Fig. 6.9 examples of SEM investigations are shown.

Figure 6.9a shows a picture of a ductile forced fracture. Such fractures occur after strong plastic deformation. In this case the coherence in the material gets lost. Figure 6.9b depicts a shrink hole in red bronze. Shrink holes or bubbles occur during solidification when pouring melts. The third picture Fig. 6.9c shows an example of stress-corrosion cracking in brass. Cracks form either along grain boundaries or across a grain in presence of a static tensile stress and simultaneous presence of corrosive materials. Brass is sensitive for stress-corrosion cracking in presence of nitric oxides, sulfur dioxide, or ammonia gas. These three pictures are courtesy of the Institut für Schadenverhütung und Schadenforschung der öffentlichen Versicherer e.V. (www.ifs-ev.org), Kiel, Germany. The last picture Fig. 6.9d shows a hole made in steel by laser radiation. This picture is courtesy of the Laserzentrum der Fachhochschule Münster, Steinfurt, Germany (Prof. B. Lödding, Prof. K. Dickmann, photograph by Dipl.- Ing. Holger Uphoff).

Also in manufacturing of semiconductors SEM is utilized to review and classify defects found on blank and patterned wafer or to investigate fine structures and patterns formed on a semiconductor in which conventional low voltage SEM (LV-SEM) has been the main image-based workhorse in critical dimensions (CD) metrology for more than 20 years. The main fields of application for fine structure investigation are after development inspection and after etch inspection. The ultimate resolution limits are in the 1.0–1.5 nm range for conventional LV-SEM.

Investigation of vertical high aspect ratio structures is a problem since for this elaborate cross-sectional sample preparation is necessary. That means to break the wafer. Then, one only sees a thin slice with perhaps inaccurate cleavage plane. To get reliable data multiple of such slices are needed which is time consuming.

6.5 Optical Coherence Tomography

Optical coherence tomography (OCT)) has developed to a proven and tested optical non-invasive imaging technology for transparent specimen since it was published first from the group of Fujimoto at the Massachussetts Institute of Technology [18] in 1991. Due to its noninvasive character OCT is widespread used in ophtalmology, cardiology, angiography, dermatology, and oncology. But it also gained interest in technical applications. It is applicable on synthetics, compound materials, glass, coatings, and ceramics to measure layer thickness, topography, warping, inner geometry, and to detect defects. A current overview on OCT can be found in [19]. OCT delivers real-time 2D cross-sectional and 3D volumetric images with micron-level resolution and millimeters of imaging depth.

The first OCT system was a *Time-Domain (TD) OCT*. The basis of TD-OCT is a Michelson interferometer or a Mach-Zehnder interferometer, a low-coherent broad-band light source, and a line-scan camera. The reference mirror is moved and creates a variable time delay of the reference beam compared to the object beam. The resulting interferogram is determined by the reflection at the surface and at each interface of two materials with different refractive index. Then, the Fourier transform

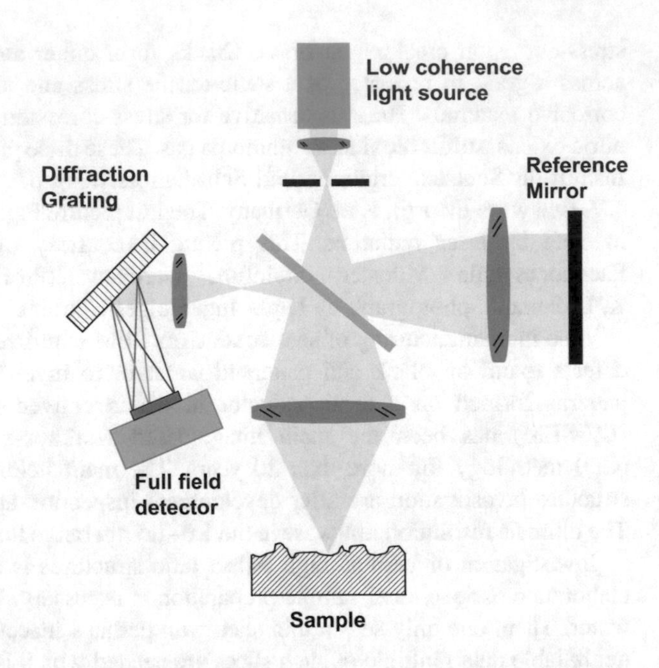

Fig. 6.10 Sketch of the Spectral Domain OCT

analysis delivers a cross-sectional image of the target in axial direction that allows evaluation of internal structures of the target material since the optical frequencies of the interferogram are related to the location in the sample where the reflection occurred. The vertical scan is called A-scan. For a complete sectional view of up to 1.5 × 1.5 mm multiple A-scans are put together in x- and y-direction to a B-scan. This technique is rather slow with about 400 A-scans per second.

The faster technique is the *Frequency Domain (FD) OCT*. Already in 1995 Fercher et al. [20] showed that with a spectral discrimination of the returning signal and Fourier analysis of the spectrum the whole sectional view can be captured at once. Then, no moving parts like the moving reference mirror in TD-OCT are required [21]. This technique is also established as *Spectral Domain (SD) OCT* or *Fourier domain OCT* where the broadband interference is discriminated with a grating and is acquired with separated detectors in a linear detector array. Measuring rates up to 85 kHz are obtained to date, allowing even *in vivo* measurements. A sketch of the SD-OCT is given in Fig. 6.10.

A second method in FD-OCT has come up in the last few years, the *Swept Source (SS) OCT*. Unless in SD-OCT here a frequency swept broadband tunable laser is used for illumination. By this way the spectral components of the interferogram are encoded in time. For decoding a balanced photodetector is used. This technique allows for up to 100,000 A-scans/s with higher penetration depth and more detailed cross-sectional views than all previous OCT platforms. A sketch of the SS-OCT is shown in Fig. 6.11.

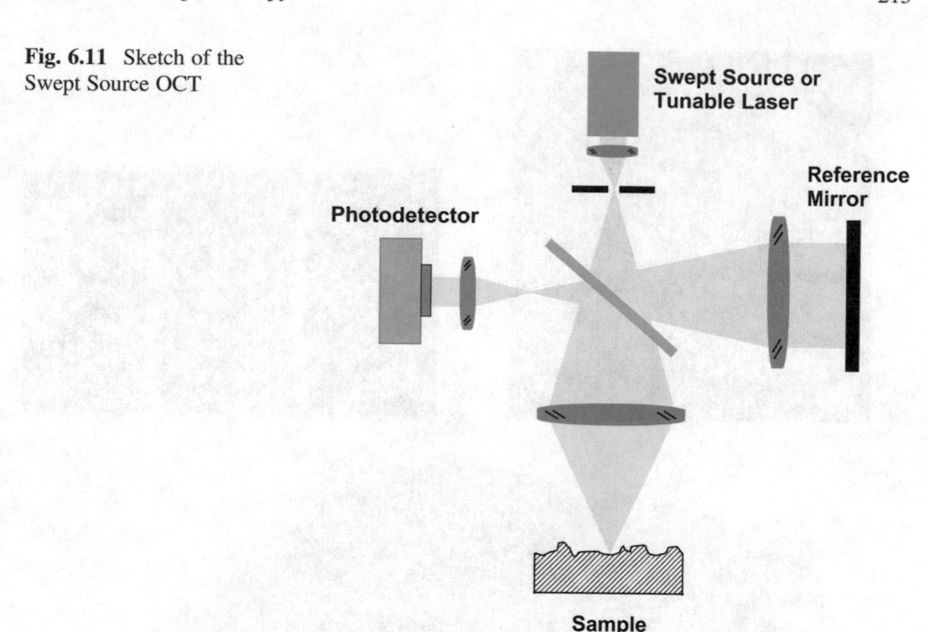

Fig. 6.11 Sketch of the Swept Source OCT

The vertical and lateral resolutions of OCT are decoupled from each another. The vertical resolution depends upon the coherence length of the light source and amounts to 1–10 μm. The lateral resolution is determined by the used optics.

The application of OCT in industrial production is still at the beginning. Nevertheless, commercial OCT systems for industrial applications like surface characterization, determination of parallelism and evenness, defect detection (particularly buried defects), or multilayer compound materials are available, e.g. from the Swiss company Flo-ir GmbH. They developed a system where an active sensor pixel array is used for the detection of light. Each of the 300 x 300 pixels together with the optics (Michelson interferometer) acts as a single interferometer with own signal preprocessing. This technique allows for a frame rate of one million frames per second. In Fig. 6.12 some results are shown which were obtained with the Flo-ir system. The pictures are courtesy of Flo-ir GmbH, Oberdorf, Switzerland.

6.6 Terahertz Spectroscopy

In *Terahertz Spectroscopy* electromagnetic waves with frequencies in the range of approximately 0.06–10 THz are used. They penetrate through foams, plastics, ceramics, and many other isolating materials since these materials are transparent in this spectral range. They cannot be applied on conducting materials or when water is present because these materials strongly reflect the terahertz radiation. Therefore,

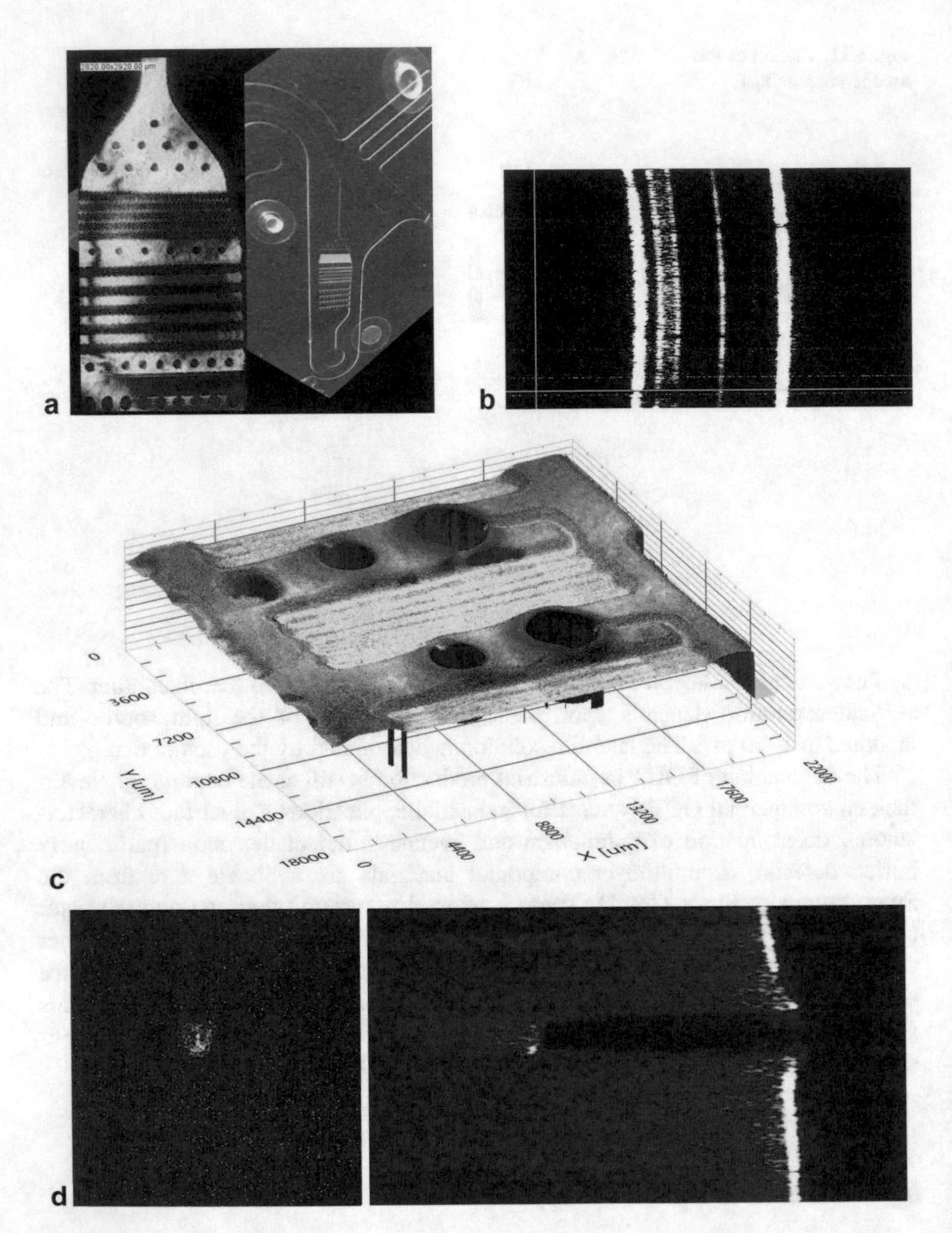

Fig. 6.12 Examples of OCT measurements: (**a**) topography of a microfluidic part, (**b**) all five layers in a transparent multilayer stack, (**c**) topography of a microfluidic channel system in polycarbonate, (**d**) a hole in a transparent film, top view and side view. (The pictures are courtesy of Flo-ir GmbH, Oberdorf, Switzerland)

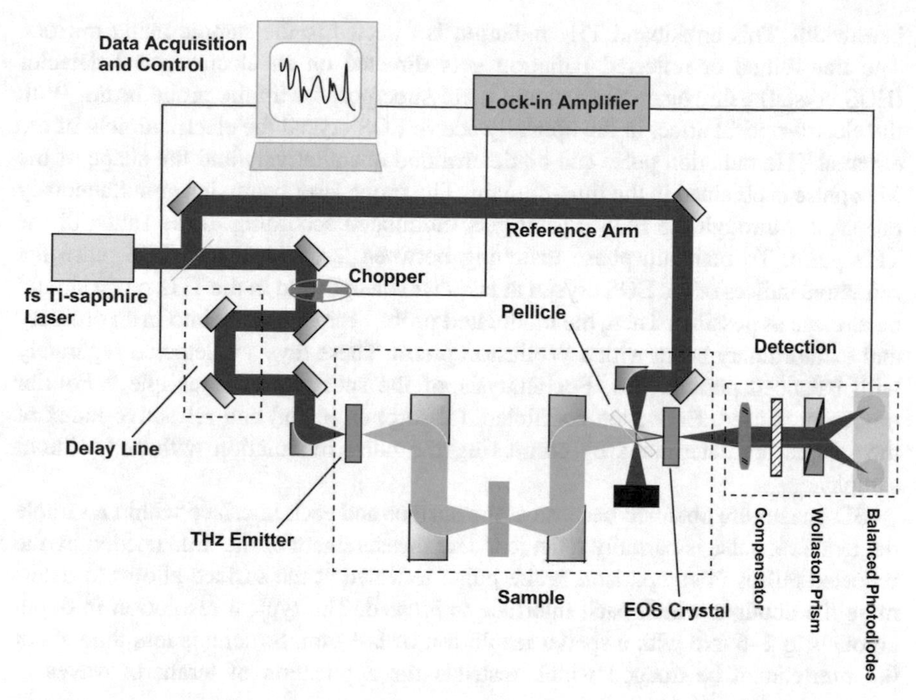

Fig. 6.13 Typical setup of THz-TDS

they do not penetrate into biological material and are completely harmless for human beings and form a well-suited alternative to X-ray inspection. As the energy of a 1 THz photon is approximately 4 meV it is too weak to break chemical bindings. Only rotations and vibrations of molecules and molecular groups can be excited. For comparison, the thermal energy at 300 K amounts to 26 meV, covalent bindings have energies of 1–10 eV, hydrogen bonds 0.5 eV, van der Waals 10–100 meV, and dipole-dipole bindings 10–30 meV.

Since the poineering works of Auston and Cheung in 1985 [22] and Fattinger and Grischkowski in 1989 [23] and 1990 [24] the *Terahertz Time Domain Spectroscopy* (THz-TDS) has evolved to the most established methods. The first application of terahertz waves goes back to Koch [25]. A review on the fundamental concepts of terahertz technology give the books from Lee [26], from Rostami, Rasooli, and Baghban [27], and from Peiponen, Zeitler, and Kuwata-Gonokami [28].

As a basic principle, the THz pulses in THz-TDS are generated by an ultra-short laser pulse and get detected coherently. The pulse length is typically less than 100 fs. Phase and amplitude of the electrical THz field are recorded at the same time. As incoherent radiation gets suppressed in this measuring technique there are no disturbances by room light or temperature. A typical setup for THz-TDS is shown in Fig. 6.13.

The laser pulse is split into a pump and a probe beam. The pump beam gets focused on a photoconductive THz emitter. It emits THz radiation with a large bandwidth. This broadband THz radiation is directed to the sample using mirrors. The transmitted or reflected radiation gets directed on an electrooptical detector (EOS crystal) using again mirrors and there superposed with the probe beam. With the electro-optical effect in the optically active EOS crystal the electrical field of the external THz radiation pulse can be determined quantitatively and the shape of the THz pulse is obtained in the time-domain. The probe laser beam that simultaneously propagates through the EOS crystal gets modulated according to the shape of the THz pulse. To maintain phase matching between laser beam and THz pulse the refractive indices of the EOS crystal at laser wavelength and in the THz range should be as close as possible. Then, the modulated probe beam gets split into in an ordinary and extraordinary beam with a Wollaston prism. These rays are detected separately with balanced photodiodes. For analysis of the spectra of the sample a Fourier transform is used. Extinction coefficient (absorption index) and refractive index of the sample are determined by comparing the pulse modulation with and without sample.

3D images are obtained because at the surface and each interface within a sample the terahertz pulse is partially reflected. The measurement of the time needed by the reflected pulses in comparison to the pulse reflected at the surface allows to determine the depth at which each interface is located. The typical resolution in depth amounts to 2–6 mm with a spatial resolution of 1–3 mm. Structures less than about 0.1 mm cannot be imaged which restricts the application of terahertz waves in topography measurement.

Terahertz spectroscopy has many applications. It is preferable for inline packaging control as well as for material inspection. Other applications are.

- Stand-off detection of hidden objects and weapons,
- Non-invasive medical and dental diagnostics
- Drug discovery and formulation analysis of coatings and cores,
- Detection of cracks and defects in solar panels
- Non-contact imaging for conservation of paintings, manuscripts and artefacts

An application where the THz technique outrivals any competitive process is the non-contact material integrity imaging of coatings and composites in automotive industry to identify the presence of defects in paints.

References

1. Antensteiner, D., Stolc, S., Pock, T.: A review of depth and normal fusion algorithms. Sensors. **18**(2), 431–455 (2018)
2. Blaschitz, B., Stolc, S., Antensteiner, D.: Geometric calibration and image rectification of a multi-line scan camera for accurate 3D reconstruction, IS&T International Symposium on Electronic Imaging 2018: Intelligent Robotics and Industrial Applications using Computer Vision 2018, Paper-Nr. 9/2018
3. Brosch, N., Stolc, S., Antensteiner, D.: Warping-based motion artifact compensation for multi-line scan light field imaging, IS&T International Symposium on Electronic Imaging 2018: Computational Imaging XVI, Paper-Nr. COIM-273

4. Horn, B.K.P.: Obtaining shape from shading information. In: Winston, P. (ed.) The Psychology of Computer Vision, pp. 115–155. McGraw-Hill, New York (1975)
5. Horn, B.K.P., Brooks, M. (eds.): Shape from Shading. The MIT Press, Cambridge, MA (1989)
6. Frankot, R.T., Chellappa, R.: A method for enforcing integrability in shape from shading algorithms. IEEE Trans. Pattern Anal. Mach. Intell. **10**, 439–451 (1988)
7. T. Wei, R. Klette, A New Algorithm for Gradient Field Integration, Image and Vision Computing New Zealand (IVCNZ'2000), Dunedin (2001)
8. Goetz, A.F.H.: Three decades of hyperspectral remote sensing of the earth: a personal view. Remote Sens. Environ. **113**(Suppl. 1), S5–S16 (2009)
9. Wolfe, W.L.: Introduction to Imaging Spectrometers. SPIE Press, Bellingham (1997)
10. Chang, C.: Hyperspectral Imaging: Techniques for Spectral Detection and Classification. Springer, Berlin (2003)
11. Grahn, H., Geladi, P.: Techniques and Applications of Hyperspectral Image Analysis. Wiley, New York (2007)
12. Knoll, M.: Aufladepotential und Sekundäremission elektronenbestrahlter Körper. Z. Tech. Phys. **16**, 467–475 (1935)
13. von Ardenne, M.: Improvements in electron microscopes, German patent GB511204 (1937)
14. von Ardenne, M., Elektronen-Rastermikroskop, D.: Theoretische Grundlagen. Z. Phys. **109** (9–10), 553–572 (1938)
15. von Ardenne, M.: Das Elektronen-Rastermikroskop. Praktische Ausführung. Z. Tech. Phys. **19**, 407–416 (1938)
16. Reimer, L.: Scanning Electron Microscopy: Physics of Image Formation and Microanalysis Springer Series in Optical Sciences, vol. 45, 2nd edn. Springer, Berlin/Heidelberg (1998)
17. Goldstein, J.I., Newbury, D.E., Michael, J.R., Ritchie, N.W.M., Scott, J.H.J., Joy, D.C.: Scanning Electron Microscopy and X-Ray Microanalysis – a Text for Biologists, Materials Scientists, and Geologists, 4th edn. Springer, New York (2018)
18. Huang, D., Swanson, E.A., Liu, C.P., Schuman, J.S., Stinson, W.G., Chang, W., Hee, M.R., Flotte, T., Gregory, K., Puliafito, C.A., Fujimoto, J.G.: Optical coherence tomography. Science. **254**(5035), 1178–1181 (1991)
19. Drexler, W., Fujimoto, J.G. (eds.): Optical coherence tomography – technology and applications, 2nd edn. Springer, Berlin (2015)
20. Fercher, A.F., Hitzenberger, C.K., Kamp, G., El-Zaiat, S.Y.: Measurement of intraocular distances by backscattering spectral interferometry. Opt. Commun. **117**, 43–48 (1995)
21. Leitgeb, R., Hitzenberger, C.K., Fercher, A.F.: Performance of Fourier domain vs. time domain optical coherence tomography. Opt. Express. **11**, 889–894 (2003)
22. Auston, D.H., Cheung, K.P.: Coherent time-domain far-infrared spectroscopy. J. Opt. Soc. Am. **B2**, 606–612 (1985)
23. Fattinger, C., Grischkowsky, D.: Appl. Phys. Lett. **54**(6), 490–492 (1989)
24. Grischkowsky, D., Keiding, S., Exter, M., Fattinger, C.: Far-infrared time-domain spectroscopy with terahertz beams of dielectrics and semiconductors. J. Opt. Soc. Am. **B7**, 2006–2015 (1990)
25. Koch, M.: Terahertz technology: a land to be discovered. Opt. Photon. News. **18**(3), 20–25 (2007)
26. Lee, Y.-S.: Principles of Terahertz Science and Technology. Springer Science+Business Media LLC, New York (2009)
27. Rostami, A., Rasooli, H., Baghban, H.: Terahertz Technology, Lecture Notes in Electrical Engineering, vol. 77. Springer, Berlin/Heidelberg (2011)
28. Peiponen, K.-E., Zeitler, A., Kuwata-Gonokami, M.: Terahertz Spectroscopy and Imaging Springer Series in Optical Sciences, vol. 171, 1st edn. Springer, Berlin/Heidelberg (2012)

Chapter 7
Multisensor – Systems – A Versatile Approach to Surface Metrology

Abstract The nature of technical relevant surfaces is often as complex as measurement of their properties cannot be carried out with a metrology tool with one or two specialized sensors. Instead, the industry increasingly looks for flexible and future-safe metrology tools that allow managing various measuring tasks or enable measuring various parameters of the same surface or workpiece.

Multisensor metrology systems might include up to four or five measurement sensors of the various technologies presented before: tactile sensors, AFM, optical point sensors, optical field of view sensors, capacitive sensors, and supplementary optical thickness sensors. The actual equipment of the multisensor metrology tool depends upon the industry branch where the tool is intended to be used. E.g., in mechanical engineering tactile sensors are combined with optical point and field of view sensors while in semiconductor industry predominately non-contact optical sensor technology, capacitive sensor technology, and SEM is used.

The main concept of a multisensor system is to have a set of sensors with each sensor at a fixed position in front of the workpiece and a x-y stage for moving the workpiece to the measurement position or to draw profiles. This is illustrated in Fig. 7.1 with a set of two chromatic white light sensors FRT CWL with different measuring range, the combination of confocal microscope and white light interferometer FRT DT, and a film thickness sensor FRT IRT800. The multiple sensor set is supported by a high-resolution camera. This picture is courtesy of FRT GmbH, Bergisch Gladbach, Germany.

To guarantee that each sensor actually measures at the desired position all sensors must be arranged exactly to each other and to the camera. With the camera the user can define the region of interest where a detailed investigation of the workpiece shall be carried out. Then, a selected sensor can be put in this region of interest to measure specific properties of the sample. The x-y stage usually consists of two single stages for each direction. For several applications these stages have an aperture to enable measurement from both sides with sensors from top and bottom. One important component in a multisensor system is also the sample holder. Often, it is customized and is designed so that measurements can be carried out in compliance with existing

M. Quinten, *A Practical Guide to Surface Metrology*, Springer Series in Measurement Science and Technology,
https://doi.org/10.1007/978-3-030-29454-0_7

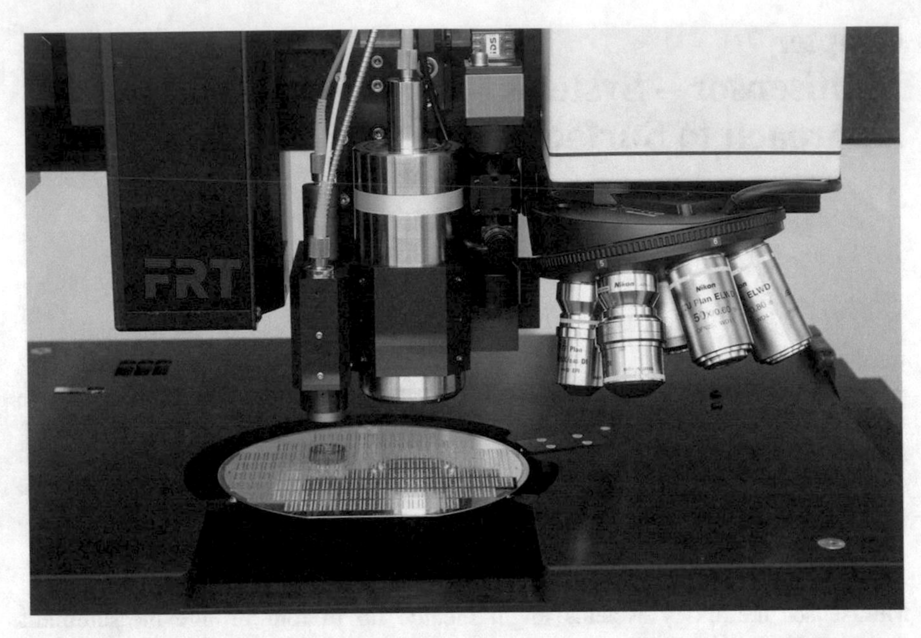

Fig. 7.1 Multiple sensors (chromatic white light, combined confocal microscope and white light interferometer, film thickness sensor) in front of a structured wafer on a x-y stage with aperture. (Courtesy of FRT GmbH, Bergisch Gladbach, Germany)

norms, e.g. a three-point wafer fixture or a fully supported wafer fixture with vacuum.

The advantage of multisensor metrology is that the strengths of the various sensors are used to measure and characterize a surface as complete as possible in a highly efficient and accurate measurement process. Further advantages are their adaptability, reduced space requirements, and the future-proof and cost-efficient operation.

The simplest multisensor system consists of two opposite sensors (laser triangulation sensors, capacitive sensors, chromatic white light sensors) for simultaneous measurement from top and bottom. It can be used to measure the thickness of opaque workpieces and is often used with capacitive and chromatic white light sensors for measurement of total thickness, total thickness variation (TTV), bow, and warp of semiconductor wafers. If in addition particular structures on this wafer shall be measured with high accuracy and resolution, an areal measurement with a confocal microscope or a white light interferometer is the preferred extension.

Another example where multisensor technology is favorable for the exact measurement of a certain quantity are samples where a structure made of an opaque material is surrounded by a transparent film, both situated on a supporting substrate. An example is a copper pillar on silicon substrate with a thin oxide film on the silicon substrate surrounding the pillar. In this system the height of the pillar is inaccesible for a single sensor if it only can measure step heights. Then, only the difference between the pillar top surface and the oxide film surface gets measured. The difference to the exact height of the pillar is the film thickness. It can be measured independently with an interferometric thickness sensor. Adding both results yields the exact height of the pillar.

Finally, all tasks can optimally be solved with multisensor systems in which a larger region of interest can easily be measured by scanning with point sensors and in which a smaller region of interest must be characterized more detailed with for example a field of view sensor.

In a modern multisensor measuring system the sensors are all integrated by metrology software. As the commonly used various sensors are fully developed the core competence lies in this software. The software of a multisensor measuring system must be able

- To define measuring tasks by apprenticeship of structures on the workpiece,
- To relate exactly the positions of the structures on the workpiece to positions of the x-y stage,
- To select the sensor that is suitable for the measuring task,
- To set sensor characteristics to values suitable for the measuring task,
- To control the selected sensor,
- To collect incoming sensor data,
- To evaluate the data for the desired quantities,
- To rate the results for the desired quantities,
- To control the x-y stage, and
- To control one or more z-stages (for all vertical scanning methods).

In fully automated measuring systems even automated sample handling and sample alignment are requested and must be controlled by the software. In addition, interfaces for transfer of data and results to the facility network must be provided. All these functions usually can be compiled in a recipe whereby the use of the system for different applications is simplified.

An example for such a fully automated system is shown in Fig. 7.2 with the MicroProf® FE. It is a multisensor capable metrology tool for front end and back end production in the semiconductor industry. The picture is courtesy of FRT GmbH, Bergisch Gladbach, Germany.

Within the housing high class (up to ISO class 3) clean room conditions prevail. The wafers are automatically taken from FOUPs/FOSBs or open cassettes at special load ports by a robot to maintain the clean room conditions inside the housing. The robot aligns the wafer and puts it on the x-y stage of the actual measuring system. The user can control and monitor the system from outside.

The term "multisensor technology" was established about 25 years ago when coordinate measuring machines (CMM) were equipped with multiple sensors with different measuring principles. Mostly, these tactile measuring systems were equipped with image processing, later also with optical sensors. The goal was to measure the once clamped workpiece completely in one measuring process and the same reference coordinate system. This task often cannot be solved with one single sensor since parts of the workpiece can either be measured only by optical means and other parts only with tactile sensors. A technique that gained more and more interest

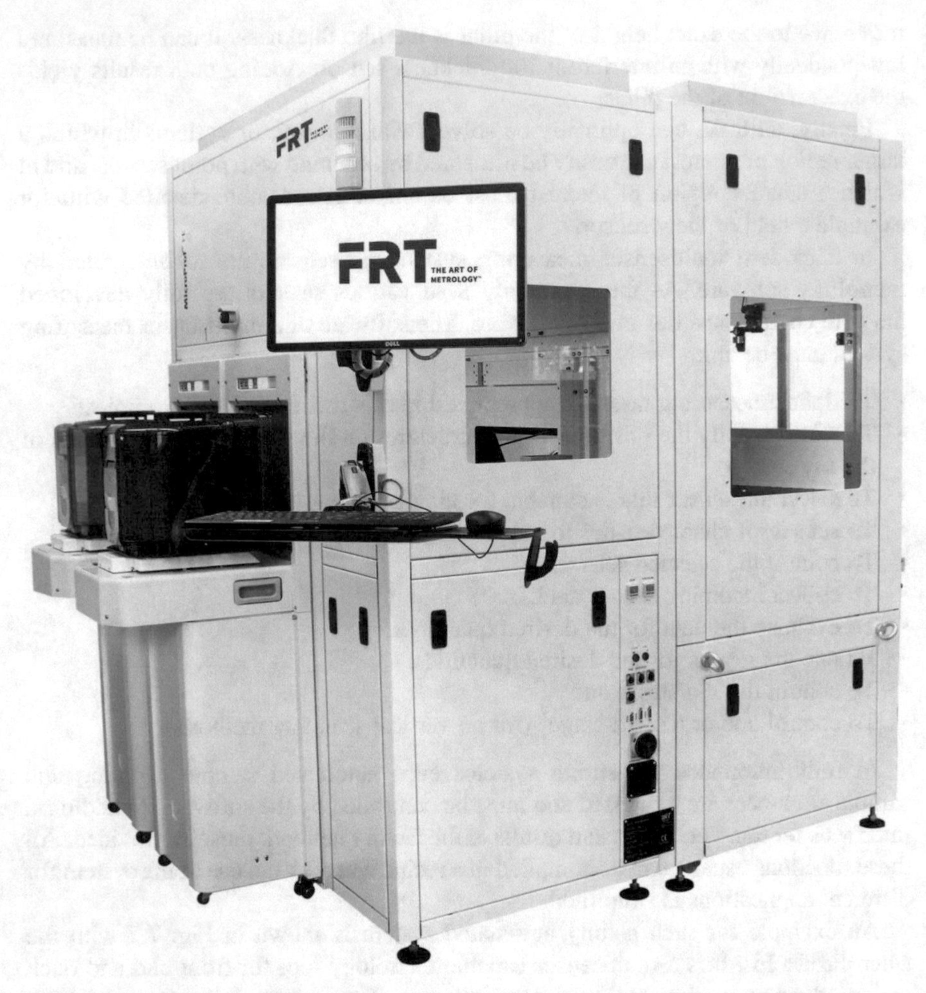

Fig. 7.2 Multisensor capable metrology tool MicroProf® FE for front end and back end production in the semiconductor industry. (Courtesy of FRT GmbH, Bergisch Gladbach, Germany)

particular in CMMs is the X-ray computer tomography. This technique delivers radiographic images under various view angles. From these images a 3D volume of the workpiece is reconstructed that gives information on the geometry of the workpiece outside *and* inside. This enables a target-performance comparison with 3D-CAD data with micrometer resolution.

Appendix – Numerics with Complex Numbers

Complex numbers are useful abstract quantities that can be used in calculations and result in physically meaningful solutions. Complex numbers have been introduced to allow for solutions of certain equations that have no real solution. For example, the quadratic equation $x^2 + 1 = 0$ has no solution in the field of real numbers. Complex numbers are a solution to this problem. The complex numbers are the field $\mathbb{C}$ of numbers of the form $z = x + i \cdot y$, where $x, y \in \mathbb{R}$, the field of real numbers, and i is the *imaginary unit* $i = \sqrt{-1}$. They extend the idea of the one-dimensional number line to the two-dimensional complex plane by using the number line for the *real part* (x-values) and adding a vertical axis for the *imaginary part* (y-values). The graphical representation of the complex number z in the complex plane is sketched in Fig. 1.

If $z = x + i \cdot y$ is a complex number, then x is called *real part* of z, i.e. $\text{Re}(z) = x$. Analogously, $\text{Im}(z) = y$ is called the *imaginary part* of z. If $z = x + i \cdot y$ is a complex number, then $z^{\sim} = x - i \cdot y$ is the complex number that lies in the conjugated plane, and is therefore called *complex conjugate number*.

In the field $\mathbb{C}$ of complex numbers there are two operations defined: "+", meaning addition, and "•", meaning multiplication.

Addition

The addition of the complex numbers z_1 and z_2 is defined as

$$z_1 + z_2 = (x_1 + i \cdot y_1) + (x_2 + i \cdot y_2) = (x_1 + x_2) + i \cdot (y_1 + y_2). \tag{1}$$

The addition is commutative, i.e. $z_1 + z_2 = z_2 + z_1$. The neutral element of the addition is $n(+) = 0 + i \cdot 0 = 0$. For the inverse element of the addition inv^+ it is $z + \text{inv}^+(z) = n(+)$, resulting in $\text{inv}^+(z) = -z$.

M. Quinten, *A Practical Guide to Surface Metrology*, Springer Series in Measurement Science and Technology,
https://doi.org/10.1007/978-3-030-29454-0

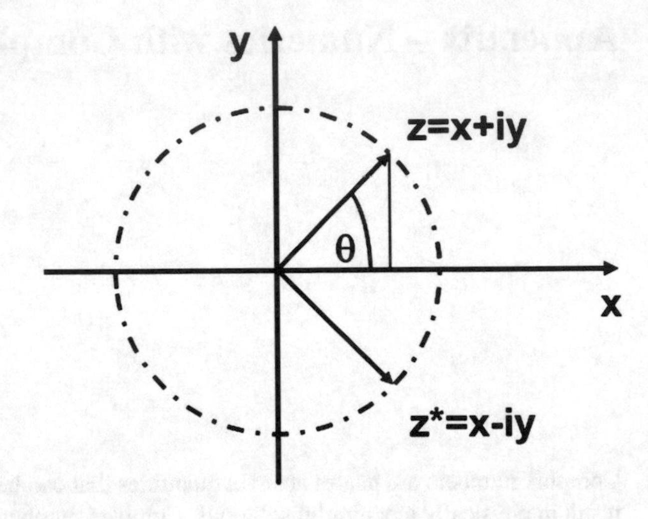

Fig. 1 Graphical representation of complex numbers

Multiplication

The multiplication of two complex numbers z_1 and z_2 is defined as

$$\begin{aligned} z_1 \cdot z_2 &= (x_1 + i \cdot y_1) \cdot (x_2 + i \cdot y_2) \\ &= (x_1 \cdot x_2 - y_1 \cdot y_2) + i \cdot (x_1 \cdot y_2 + x_2 \cdot y_1) \end{aligned} \tag{2}$$

The multiplication is commutative, i.e. $z_1 \cdot z_2 = z_1 \cdot z_2$. The neutral element of the multiplication is $n(\bullet) = 1 + i \cdot 0 = 1$. For the inverse element of the multiplication $inv^{\bullet}$ it is $z \cdot inv^{\bullet}(z) = n(\bullet)$, resulting in $inv^{\bullet}(z) = 1/z$ for all complex numbers $\neq 0$. With the help of the complex conjugate number $z^{\sim}$ it can be expressed as.

$$inv^{\bullet}(z) = 1/z = z^{\sim}/(z \cdot z^{\sim}) \tag{3}$$

Modulus

The modulus of a complex number $z = x + i \cdot y$ corresponds to the length of the pointer in Fig. 1. It is the hypotenuse of the triangle formed by the real part x and the imaginary part y as legs of a right-angled triangle. Therefore, the modulus $|z|$ follows as

$$|z| = \sqrt{x^2 + y^2}. \tag{4}$$

For all complex numbers with the same modulus the corresponding pointer ends on the dash-dotted circle in Fig. 1. From this graphical representation the relation

$$z = x + i \cdot y = |z| \cdot (\cos(\theta) + i \cdot \sin(\theta)) = |z| \cdot \exp(i\theta) \tag{5}$$

can be deduced with $\theta = \arg(z) = \tan^{-1}(y/x)$ being the *argument* of z. This is the polar representation of a complex number. The analytical identity $\cos(\theta) + i \cdot \sin(\theta) = \exp(i \cdot \theta)$ allows for the application of power laws when calculating with complex numbers. The multiplication of z with its complex conjugate number $z^{\sim}$ yields $|z|^2 = z \cdot z^{\sim}$.

Division

The division of two complex numbers z_1/z_2 can be reformulated into a multiplication of two complex numbers z_1 and $(1/z_2) = \text{inv}^{\bullet}(z_2)$. Hence, the division is defined as

$$\frac{z_1}{z_2} = z_1 \cdot \text{inv}^{\bullet}(z_2) = \frac{z_1 \cdot \tilde{z_2}}{z_2 \tilde{z_2}} = \frac{z_1 \cdot \tilde{z_2}}{|z_2|^2} \tag{6}$$

Power n

To calculate z^n with n being a real number $n \in \mathbb{R}$, the polar representation of a complex number Eq. (5) is useful. Then

$$z^n = |z|^n \cdot (\cos(n\theta) + i \cdot \sin(n\theta)) = |z|^n \cdot \exp(in\theta) \tag{7}$$

For $n \in \mathbb{N}$, i.e. a positive integer number, z^n is given by

$$\begin{aligned} z^n = {} & \left[x^n - \binom{n}{2} \cdot x^{n-2}y^2 + \binom{n}{4} \cdot x^{n-4}y^4 - \ldots\right] \\ & + i \cdot \left[\binom{n}{1} \cdot x^{n-1}y - \binom{n}{3} \cdot x^{n-3}y^3 + \ldots\right]. \end{aligned} \tag{8}$$

Logarithm

The natural logarithm log(z) of a complex number z can easily be calculated using again the polar representation Eq. (5):

$$\log(z) = \log(|z| \cdot \exp(i\theta)) = \log(|z|) + i \cdot \theta. \tag{9}$$

Exponentiation

For the complex exponentiation ${z_1}^{z_2}$ the exponential function can be used:

$${z_1}^{z_2} = (\exp(\log(z_1)))^{z_2} = \exp(z_2 \cdot (\log(|z_1|) + i\theta_1)). \tag{10}$$

Trigonometric Functions

The trigonometric functions sin(z) and cos(z) can be calculated using the analytical identity $\cos(z) + i{\cdot}\sin(z) = \exp(iz)$. Then

$$\sin(z) = \frac{\exp(iz) - \exp(-iz)}{2 \cdot i} = \sin(x) \cdot \cosh(y) + i \cdot \cos(x) \cdot \sinh(y) \tag{11}$$

$$\cos(z) = \frac{\exp(iz) + \exp(-iz)}{2} = \cos(x) \cdot \cosh(y) - i \cdot \sin(x) \cdot \sinh(y) \tag{12}$$

Index

A
Abbe criterion, 10
Abott-Firestone curve, 24
Absorption index, 75
Accuracy, 10
Angular resolved scattering (ARS), 89, 169
Anisotropy, 76
Anomal dispersion, 77
Arithmetic mean deviation, 20
Atomic force microscope (AFM), 47
Autocorrelation function, 24
Autofocus sensor, 102
Average peak to valley height, 23

B
Bearing ratio, 24
Bidirectional reflection distribution function (BRDF), 89
Bidirectional scatter distribution function (BSDF), 89
Brendel oscillator, 79
Bruggeman effective medium, 187

C
Cantilever, 48
Capacitance measurement, 57
Cauchy formula, 80
Charged coupled device (CCD), 36
Chemical force microscopy (CFM), 53
Chromatic aberration, 85, 95
Chromatic white light sensor, 95
Coherence, 74
Coherence length, 74
Combined standard uncertainty, 13
Complementary metal oxide semiconductor (CMOS), 36
Computer-generated holograms (CGH), 133
Confocal microscope, 115
Confocal point sensor, 115
Conoscopy, 159
Contact mode, 50
Contrast, 112
Critical angle of total reflection, 82
Critical dimensions, 186
Critical gage capability index, 14
Cut-off wavelength, 18

D
Dark current, 35
Definition region, 29
Deflectometry, 163
Depth of field (DoF), 111
Depth of focus, 112
Dielectric function, 75
Differential interference contrast, 113
Diffraction, 86
Diffractive grazing-incidence interferometer, 151
Digital holographic microscopy (DHM), 155
Digital holography, 155
Dispersion, 84
Dispersion integrals, 79
Downhill simplex algorithm, 183
Drude susceptibility, 78

M. Quinten, *A Practical Guide to Surface Metrology*, Springer Series in Measurement Science and Technology,
https://doi.org/10.1007/978-3-030-29454-0

E
Eddy currents, 62
Electromagnetic radiation, 67
Ellipsometric parameters, 188
Ellipsometry, 188
Encrgy-dispersive X ray spectroscopy (EDX), 209
Evaluation length, 18
Evaluation region, 29
Evanescent wave, 82
Expanded uncertainty, 13
Exponential Cauchy formula, 80
Extinction coefficient, 75
Extraordinary ray, 76

F
Fast Fourier transform (FFT), 182
F-filter, 29
Fluorescence microscopy, 112
Focal depth variation, 124
Focus variation, 124
F-operator, 29
Force modulation microscopy, 52
Form deviations, 7
Form filter, 18
Four pi microscopy, 121
Frequency-domain OCT (FD-OCT), 212
Frequency scanning interferometry (FSI), 153
Fresnel coefficients, 83
Fresnel equations, 83
Fringe projection, 106

G
Gage repeatability & reproducibility, 15
Grating equation, 88
Grating function (GF), 87
Grating period, 87
Grazing incidence interferometry (GII), 150
Groove density, 87

H
Harmonic oscillator, 76
Harmonic wave, 69
Holography, 155
Huygens-Fresnel principle, 71
Hyperspectral imaging (HSI), 205

I
Industrial image processing (IIP), 199
Interference, 72
International Standardization Organization (ISO), 15

K
Kelvin probe microscopy, 53
Kim oscillator, 79
Kramers-Kronig relations, 79
Kurtosis, 22

L
Laser point triangulator, 104
Laser scanning confocal microscope (LSCM), 118
Lateral force microscopy (LFM), 53
Levenberg-Marquardt algorithm, 183
L-filter, 30
Light, 67
Light scattering, 167
Linear superposition, 72
Line projection, 105
Linnik, 141
Lorentz oscillator, 76
Low coherence interferometry, 136

M
Machine vision, 199
Makyoh topography sensor, 165
Maximum peak to valley height, 23
Maximum profile peak height, 22
Maximum profile valley depth, 22
Maxwell-Garnett effective medium, 187
Measured surface, 6
Metal oxide semiconductor field effect transistor (MOSFET), 35
Microscope, 108
Mirau, 140
Model based infrared reflectometry (MBIR), 187
Multi-wavelength interferometry (MWLI), 147

N
Nesting index, 29
Nipkow disk, 120

Noise filter, 18
Nomarski prism, 113
Nominal surface, 5
Non-contact mode, 50
Nonlinear regression analysis, 182
Normal dispersion, 77
Numerical aperture, 110

O
Optical coherence tomography (OCT), 211
Ordinary ray, 76

P
Partially developed speckles, 175
Phase contrast microscopy, 112
Phase shifting interferometry (PSI), 129
Point-spread function (PSF), 115
Polarization, 71
Potential gage capability index, 14
Power spectral distribution (PSD), 90
Primary parameters, 18
Primary profile, 18
Pushbroom imagers, 206

R
Rayleigh criterion, 10
Real surface, 6
Reflection law, 81
Reflectometric measurement, 177
Reflectometry, 177
Refractive index, 75
Regression analysis, 183
Repeatability, 12
Reproducibility, 12
Resolution, 10
Root mean square deviation, 21
Roughness, 7
Roughness filter, 18

S
Sampling length, 18
Scanning electron microscopy (SEM), 208
Scanning near-field optical microscope (SNOM), 127
Scanning surface potential microscopy, 53
Scattering, 88
Schott formula, 80
Sellmeier formula, 79
S-filter, 29
S-F surface, 29, 30
Shack-Hartmann sensor, 160
Shape from focus, 124
Shape from shading (SfS), 203
Signal-to-noise ratio (SNR), 38, 180
Skewness, 22
S-L surface, 29, 30
Snell's law of refraction, 82
Sparrow criterion, 10
Spatial scanning HSI, 206
Speckle, 167, 172
Speckle contrast, 174
Speckle correlation, 174
Spectral domain OCT (SD-OCT), 212
Spectral scanning HSI, 206
Stability, 14
Standard deviation, 12
Staring imagers, 206
Stimulated emission depletion, 122
Stylus tip, 43
Swept source OCT (SS-OCT), 212

T
Tactile surface profiling, 43
Tapping mode, 51
Tauc-Lorentz oscillator, 79
Terahertz spectroscopy, 213
Terahertz time-domain spectroscopy (THz-TDS), 215
Through-focus scanning optical microscopy (TSOM), 124
Tilted wave interferometry (TWI), 134
Time-domain OCT (TD-OCT), 211
Tolerance, 14
Topography measurement, 7
Total height of profile, 22
Total integrated scattering (TIS), 90, 168
Total reflection, 82
Traceability, 14
Tracing length, 18
Transfer characteristic, 19

U
Uncertainty, 12
Useful dynamic range, 38

V

van der Waals force, 51

W

Wave, 67
Wave front sensing, 160
Wavelength scanning interferometry (WSI), 145
Waviness, 7
White light interferometry (WLI), 136
White light LED, 38

Z

Zernike polynomials, 134

Printed in the United States
by Baker & Taylor Publisher Services